COPYRIGHT

ISBN: 9781655583711

JCOM Automation Inc.
Peterborough, Ontario, Canada
jcomautomation.ca

Note:
HART and WirelessHART are registered trademarks of the FieldComm Group Inc., Austin, Texas, USA. AMS Trex Device Communicator is a trademark of Emerson. SIMATIC is a trademark of Siemens AG. PACTware is a trademark of the PACTware Consortium e. V.

CONTENTS

To my Dad, A. Edward Powell,
a practical engineer and one of
my biggest supporters.

ACKNOWLEDGMENTS

To quote a friend who paraphrased an old saying: "It takes a village to write a book."

This village started with Henry Vandelinde, Ph.D., who taught me to write and was my co-author for *Catching the Process Fieldbus*. Thank you Henry—I can still see you pointing to a paragraph and asking me, "What on earth does this mean, Powell?"

This project was initiated and supported by Dana Van Allen of CloseTalk Communications, who made the suggestion about working with her as an editor and designer for a small ebook. I don't think you knew what you were starting when you asked that question. Thank you, Dana!

The seeds for this project and the ones that will follow were planted by Pieter Barendrecht of PROCENTEC who suggested that my writing days were not—and should not—be over.

I want to thank the following companies who provided pictures and let me use them in this guide: Vega, Thorsis Technology GmbH, Helmholz, PROCENTEC, and Siemens.

My reviewers, Mark Wheeler, Russ Muller, Kyle Roos, Peter Thomas, Henry Vandelinde, and John Rinaldi, were excellent help. Thank you all—and in particular Russ, for encouraging me to expand this into the form that it is now.

I was surprised when Thomas J. Burke offered to write the foreward and even more surprised when I read it. Thank you so much for the glowing review and a great foreward.

Last but not least, thank you to my wife Debbie Challner and daughters Maya Powell and Luannah Powell for their support during this project.

FOREWARD

I've had the unique opportunity to write the foreward for this very electrifying book.

From first glance I found the book extremely informative, and very delightful to read. I realize this book will be a prodigious valuable reference to continue to understand the value of and how to maximize the effectiveness of using the HART communication protocol.

Over the years, I've had multiple opportunities developing hardware/ software and working closely with end users to build solutions out of many disconnected components. I wish I would've had a book like this so many times to help me connect all the disconnected pieces together and to make them work in a highly reliable, effective way that maximized operations.

In the world today, we are overwhelmed with technology innovations and everyone talking about all the three- and four-letter acronyms across international boundaries. Data and information integration interoperability are so very important to achieve with all the multitude of connected devices. Of course, there are many opportunities to connect up and exchange data between the multitude of disparate devices and applications.

The challenges that face the end users, owner operators, and suppliers, are how to manage these devices and applications and to establish and maximize an effective strategy to build effective solutions.

Understanding and maximizing how to use the various field bus protocols is key to getting the data and to transform the data into useful information from the connected devices and potential connected devices. Getting the data is the foundation to truly maximize efficiency of any operation in process automation and beyond.

I've had the opportunity to work with Industry 4.0 and regional equivalents, as well as working with a multitude of standards organizations and, most importantly, all the end users and what we consider the owner operators across such industries as oil and gas, chemical, water treatment, and pharmaceuticals, just to name a few. They all depend on connecting all the disparate applications and devices to get the data necessary, and one of most important communication protocols clearly is the HART communication protocol for process automation.

When you think about building bi-directional communications schemes, security and safety are so absolutely required for communication between control system components inclusive of DCS, PLC and all the IO devices out there, including sensors and actuators. In a complex world, it's very important to have reliability, security redundancy, and safety built into the equation between all the many devices that make up a company's process automation system.

In the early days, you couldn't even imagine talking about industry-standard communication protocols in process automation. Everything was based on what I call proprietary open systems that single vendors promoted as the solution for connectivity. The problem was you were locked into a single vendor's solution and the set of vendors that they wanted you to be connected to. The HART communication protocol truly is a means to an end, providing the necessary wired and wireless connectivity to connect process automation devices, enabling connected solutions.

Suppliers, and most importantly end users, across the industrial automation landscape are looking for solutions to problems they have today and will have tomorrow, and they need to have a good understanding of how to maximize the use of very important technologies like the HART communication protocol.

James has written a very informative book that will allow you to understand the value proposition of HART, where it came from and where it's going, and how to truly leverage and be able to successfully use products that are connected via the HART communication protocol.

James' background is a true testimonial and he excels in his knowledge and understanding of the value and how to get the most out of any communication protocol; James is known for helping suppliers and end users build complete connected solutions. James understands the value of data and how to get the data via such established communication protocols like the HART protocol.

He understands the technology and he communicates the important things that you as the reader must know in order to really maximize your operation based on the connectivity to the different devices.

James' informative background allows him to navigate around the potential pitfalls that you could walk into without this informative book.

The most important thing associated with any field bus and communication protocol is understanding how to use them successfully. There is a lot more than just buying products from different vendors and installing them. Supplier and end users need to understand the what and how of installation and configuration in such a way that the devices and systems work together to maximum efficiency.

James has a longstanding history with this and he is one of the key players in the process control industry, understanding all the nuances of the myriad of devices and products. The challenges of configuring the various devices and getting them to work is easy once you understand what James has described in his very effective book.

I really wish to congratulate you as the reader on the informative decision that you have made picking up this book and using it as your guide to truly understanding the HART communication protocol. As you read this book, pay close attention to the very important details and guidelines that James has provided you to truly maximize your usage of the technology.

It is of great pleasure that I've had the good fortune to know James.

Thomas J. Burke
Visionary and OPC Foundation Founder

CHAPTER 1: INTRODUCTION

"Begin at the beginning," the King said gravely, "and go on till you come to the end: then stop."

--Lewis Carroll

When I began my career in Process Automation over 20 years ago, Highway Addressable Remote Transducer (HART) was old.

Like Modbus, it was being challenged by all the 'new' protocols in town, and many forecasted its demise. Truth be told, I was one of them. Now more than two decades later, HART is still going strong; in fact, there is a good chance it will even outlive some of those 'new' protocols that were going to replace it.

Why? Good question—I suspect it is because HART is simple and it works. Now, as we will see, it has very easy and cost-effective connections to Industrial Ethernet, the backbone of IIoT.

So, as Alice in Wonderland's king said, let's start at the beginning. However, with this protocol, the end is nowhere in sight.

HART: A BRIEF HISTORY

HART was developed in the mid to late 1980s by Rosemount, as a digital replacement for 4-20 mA technology for transmitting a process variable for analog instruments. Its design mandate was to use the same wiring as 4-20 mA and provide a path forward for more advanced features that would only be possible with two-way digital communications.

Later on, some considered the protocol as a transition from the 4-20 mA technology that dominated the industry to the fully digital protocols like PROFIBUS PA and Foundation Fieldbus. HART was first proprietary and then was soon made 'open' and available to all instrument vendors.

Rosemount, which is now part of Emerson, was the initial designer of the protocol. They then made it open by donating the intellectual property to the HART Communication Foundation, an open organization whose sole purpose was to promote and develop the HART protocol.

This group then eventually merged with the Fieldbus Foundation to form the FieldComm Group. The FieldComm Group are the current owners of this open protocol.

HART's design goals were to:

- Use the same wires as 4-20 mA
- Allow for remote configuration
- Allow for remote diagnostics

- Be able to transmit multiple variables
- Be intrinsically safe
- Have the ability to be multi-dropped

HART was able to do this by imposing a digital signal over the standard 4-20 mA signal using Frequency Shift Keying (FSK) technology. The data rate is slow, running at 1200 bits per second—which translates into two to three updates per second. However, for most analog applications, this is still fast enough.

Most older HART installations still use the analog channel (aka 4-20 mA) to transmit the main process variable and then use the HART signal for remote configuration and diagnostics. The use of 4-20 mA to transfer the process variable is changing as industry realizes the values of the additional accuracy and diagnostics the digital channel provides.

The design of HART was very successful, and today, most instruments support this open protocol. However, even though HART is available in instruments throughout industry, most users do not have a good understanding of the protocol—and, more importantly, how to use it properly.

Since the protocol's initial release, it has grown in scope and functionality. Now there are multiple physical layers and expanded functionality: WirelessHART and HART-IP.

The purpose of this guide is to look at the protocol from a practical point of view and show how to use it properly. As we'll see, when used within practical limits, the protocol is very effective.

◆ ◆ ◆

Every protocol has its limits!

In working with many different protocols over the years, I have noticed that every protocol has its place—applications where it really excels. HART is no different.

As we will see in this guide, HART used in point-to-point topology, where the run is under 100 metres, is excellent.

The communications work with no errors, and the speed is acceptable. The protocol is simple to implement and can easily be integrated into any instrument. It gets the job done.

This guide will also talk about areas where, yes, you can use HART, but really it is not the best idea. Of course, it is up to the end user to pick wisely.

HART PHYSICAL LAYERS/ TYPES OF HART

HART has developed several physical layers:

- FSK (serial HART/standard HART)
- C8PSK (coherent 8-way phase shift keying)
- HART-IP
- WirelessHART

Frequency Shift Keying (FSK) was the original physical layer and still the most popular. It runs at 1200 baud and is often called standard HART or serial HART.

Coherent 8-Way Phase Shift Keying (C8PSK) is in the HART standard and is eight times faster than standard HART. When it was first proposed back in 2001, it was technically very hard to do.

However, that has changed, and now with the emphasis on Industry 4.0, there is some activity in the marketplace to get this option implemented. Unfortunately, it is not used in any practical sense at the time of this publication.

HART-IP is basically the HART message encapsulated into an TCP/IP or UDP/IP message. It is used extensively to communicate between engineering stations or controller and HART smart cards.

WirelessHART is part of the HART V7 and is widely used. WirelessHART uses a wireless mesh network to communicate to a network of WirelessHART devices. It can integrate standard HART devices seamlessly.

ADDITIONAL HART FUNCTIONS

Burst mode:

Burst mode is a function that some HART slaves support where the master can request the slave to continuously send out its process variables in a message. This increases the update rate from two updates per second on point-to-point to three to four updates per second.

Typically, the slave would need to have burst mode enabled, and then the master would send a message to the slave for the slave to enter burst mode. The master can end this mode as well.

Although this is a neat feature to increase data rate, it is not used much, and many instruments do not even support it.

Configuration change flag/Change flag

Whenever a change is made to a configuration parameter in a HART device, the 'configuration change flag' (also sometimes called 'change flag') is set.

The idea is that after an end-user has set up a HART device completely, they need to clear this change flag. Most configuration software and handhelds have a function to do this. For this function to work properly, the user has to be diligent in clearing this flag after every configuration change.

When troubleshooting an instrument, knowing if a configuration change had occurred is very useful. By having HART devices track this with the configuration change flag, this information can be recorded in the automation system. This information is then available to the maintenance staff.

HART modem

Most HART devices use the FSK physical layer and are typically wired into a 4-20 mA input card in a PLC. To use HART in these cases, you need to use a HART modem—and, as we will learn later, have a HART resistor. There are many modems on the market today made by a several companies: Thorsis Technologies GmbH, Pepperl+Fuchs, Phoenix Contact, to name a few.

When choosing a modem, it is best to verify that it is supported by the software that you want to use—for example, both Thorsis Technologies GmbH and Pepperl+Fuch support Siemens PDM and PACTware software.

Most computers today have USB interfaces. In the past, there were RS-232 ports on laptops, and there are RS-232 to HART modems available. They work fine, but please note that these cannot be combined with an RS-232 to USB converter. There are known timing issues that cause communication problems.

Integrating HART devices into the control system

As already noted, many HART devices have been wired into 4-20 mA cards. Using 4-20 mA cards means that the control system cannot gain any benefits from HART; instead, HART is only used for initial setup and maybe maintenance.

The real value of HART is only realized when a HART instrument is wired into a HART smart card. These smart cards typically can be set up to monitor the 4-20 mA channel and communicate to the device via the HART protocol. These cards are either located in a remote I/O rack or as part of a gateway device. Pictured below is the Modbus TCP or PROFINET to HART gateway by Thorsis Technologies GmbH.

When HART smart cards are used, it is then possible to communicate to the HART instruments over the network, and it is also possible to get status information.

Both offer tremendous operational benefits. Once HART is networked, then your control system can react to a bad signal and your maintenance staff can see what is going on.

Use of these HART smart cards is the future.

Conclusion

HART smart cards are one reason why HART FSK is still going strong. Another, and more important, reason is how easy it is for device vendors to integrate it into their products. The additional hardware costs to add HART to a 4-20 mA device is pennies. Development still costs, but having the shelf cost so low makes it easier for device vendors.

The next part of our HART journey is to dive into the protocol itself, and in doing so, put HART status information into perspective.

CHAPTER 2: HART PROTOCOL AND STATUSES

"Anyone who considers protocol unimportant has never dealt with a cat."
--Robert A. Heinlein

The HART protocol is one of the easiest protocols in automation. Modbus is easier, but not by much. This has helped a lot with HART's adoption.

This chapter takes a look at the protocol for two reasons: first, when you use HART, sometimes errors will occur in communicating.

If you know nothing of the protocol, then you will not know if you should be alarmed or not.

Second, one of the most important features of HART is its ability to tell the end-user if there is a problem. HART does this by providing status information.

However, this 'status' information was not designed up front. It has grown over time and has the personality of a cat. If you do not approach it cautiously, it will bite you. You need to get to know the protocol, maybe scratch its ears first. For HART, the commands are the keys to understanding all the different statuses.

The HART protocol itself is a simple master/slave type protocol with the ability to have two masters (one primary and one secondary). It is basically the same protocol on all the different physical layers. Many of the physically layers permit several slaves on one network.

◆ ◆ ◆

Demystifying the magic

I remember as a kid thinking how neat it was to see a car drive down the road—how did it do that? Fast forward to high school shop class: the teacher popped open the hood, revealing "the magic," except now it was no longer magic.

This is why I like to show details of the actual packets and how the protocol works: it takes away the magic. HART, as we will see in the following section, is pretty simple but well designed. Pay particular attention to the classifications of the commands.

HART PACKET

The HART packet structure is simple:

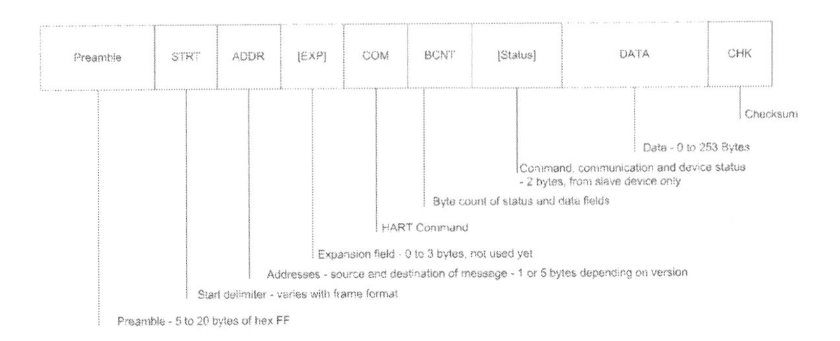

Example:

Master to Slave (Command HART V5.0+)

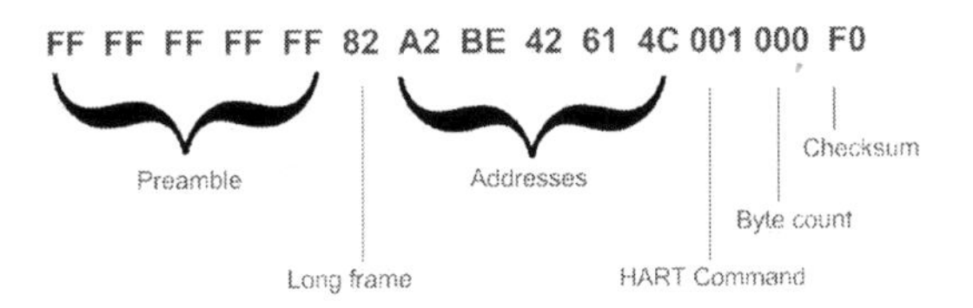

Preamble, start delimiter, source and destination address, command, byte count, status, data, checksum: pretty much the basics of any communication protocol.

Please note that WirelessHART is different in its message structure. It completely changes the first part of the message but keeps the HART command structure. This is why the address range and setup is different, while the functionality is still standard HART.

ADDRESSING

The addressing method in HART uses one address for the slaves and a second address (a bit field) for the master.

The primary master is at address 1, and the secondary master is at address 0.

The slaves use address 0 to 15 for HART V5 and below and 0 to 63 for HART V6 and above.

This address is referred to as the short frame address.

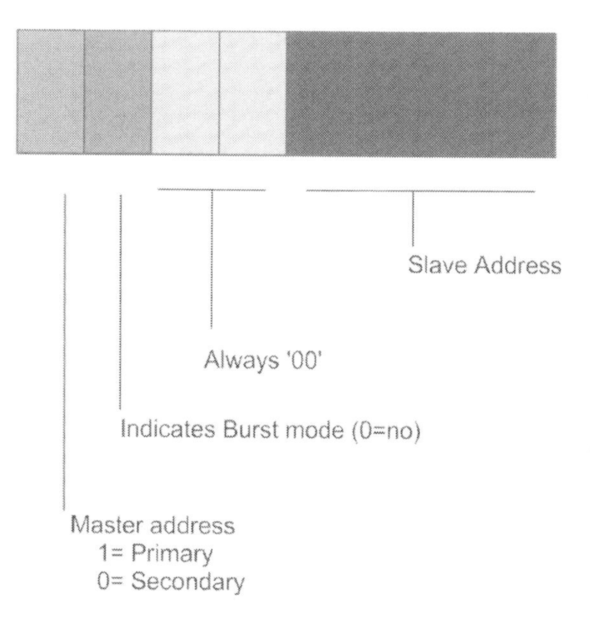

V5 of the protocol introduced the use of a unique identifier as the address in all communications except for the first one.

The first message uses a short message format and the serial address to find out the unique identifier of the device. Then all additional messages use the long message format and the unique identifier to address the device. The reason they did this was to make certain that they were talking to the correct device.

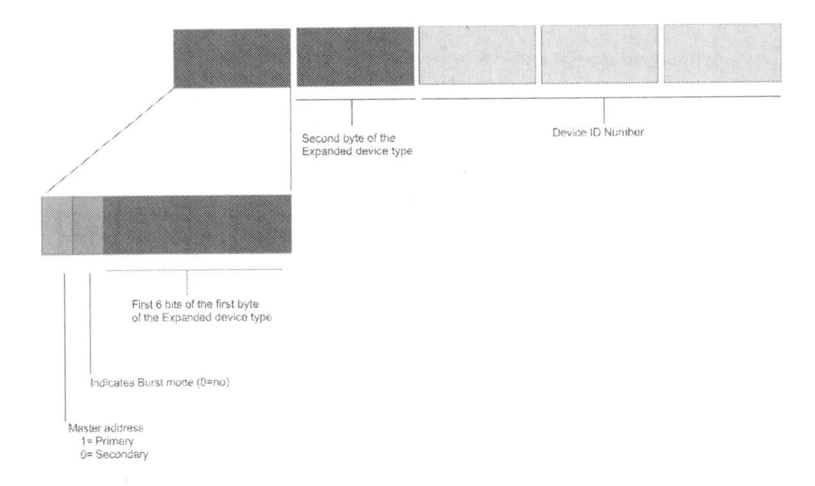

Most HART installations are set up for point-to-point configuration—with one master and one slave. However, you can also use HART multi-drop. Both of these topologies are discussed later in this guide.

COMMANDS

The beauty and simplicity of HART is how it defines its commands. The commands are divided up into three classifications:

- Universal commands, which are commands 0 to 30. These are commands that every HART slave must support. In V7 of the protocol, commands 38 and 48 were added to the universal commands.
- Common practice commands are commands 32 to 121. These are commands that are common to many types of devices, which the device manufacturer has the option of using or not.
- Device-specific commands are commands 128 to 253. These are commands that are fully defined by the device manufacturer.

It is important knowing the different types of commands when you use any configuration or diagnostic equipment.

Configuration and diagnostic equipment will give you the option of either using a device-specific driver or a generic driver. If you use the generic driver, then only universal and common practice commands are used.

However, as we learnt above, not all common practice commands have to be implemented in a device. Therefore, errors can be expected and are not a problem.

Errors are not always bad

I remember the first time I connected to an instrument using a 'generic' Device Description (DD), I was able to read the tag name and value, but I kept getting all these errors saying "unsupported command."

Why? What was going on?

After learning more about the protocol, I realized what I was seeing was that the information I was getting back were from the universal commands—and all the errors were coming from the common practice commands that were not implemented in that instrument.

I could still use the 'generic' DD, but I had to accept the fact that, depending on the instrument that I was connecting to and what I was trying to do, errors were simply going to happen.

These errors were not a problem, but just reflected what commands were implemented in the device.

HART V5 START-UP SEQUENCE

When a master first talks to a device using HART V5 and above, it needs to get the unique identifier to be able to talk to it with most commands. It does this by first issuing a universal command 0: *read unique identifier*. The data returned by this command is:

- Byte 0: 'FE' (expansion)
- Byte 1-2: Expanded device type
- Byte 3: Minimum number of preambles required
- Byte 4: Universal command revision
- Byte 5: Device-specific command revision
- Byte 6: Software revision
- Byte 7: Hardware revision
- Byte 8: Device function flags
- Byte 9-11: Device ID number
- Byte 12: Minimum number of preambles from slave
- Byte 13: Maximum number of device variables
- Byte 14-15: Configuration change counter
- Byte 16: Extended field device status
- Byte 17-18: Manufacturer identification code
- Byte 19-20: Private label distributor code
- Byte 21: Device profile

Once the master has this information, it can then use the command that it wanted to in the first place. The unique identifier is formed from information in the expanded device type and the Device ID number.

Below is an example of a Thorsis Technologies isNet HART gateway talking to a field device. On power-up, the gateway issues a series of commands 0 and waits for a response from the slave.

Once the slave has powered up, it responds to command 0. Then the master can form the unique identifier—and in this case, use command 3 to read the current value of the process variable and up to four dynamic variables.

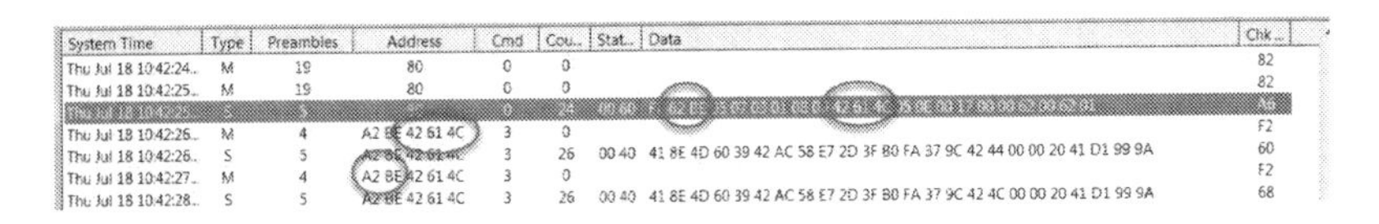

System Time	Type	Preambles	Address	Cmd	Cou..	Stat..	Data	Chk ..
Thu Jul 18 10:42:24..	M	19	80	0	0			82
Thu Jul 18 10:42:25..	M	19	80	0	0			82
[illegible]	S	5	[illegible]	0	24	00 60	[illegible]	A6
Thu Jul 18 10:42:26..	M	4	A2 BE 42 61 4C	3	0			F2
Thu Jul 18 10:42:26..	S	5	A2 BE 42 61 4C	3	26	00 40	41 8E 4D 60 39 42 AC 58 E7 2D 3F B0 FA 37 9C 42 44 00 00 20 41 D1 99 9A	60
Thu Jul 18 10:42:27..	M	4	A2 BE 42 61 4C	3	0			F2
Thu Jul 18 10:42:28..	S	5	A2 BE 42 61 4C	3	26	00 40	41 8E 4D 60 39 42 AC 58 E7 2D 3F B0 FA 37 9C 42 4C 00 00 20 41 D1 99 9A	68

HART VERSIONS

HART has had a busy history since its development in the early 1980s. On the following page is a summary of the different versions of HART. HART designers have been committed to making the revisions both backward and forward compatible. This is true in theory and almost true in practice.

Very few HART devices made to V1 through V4 are currently active in the field. The majority of devices in operation are written to V5, with many new devices now being written to V7.

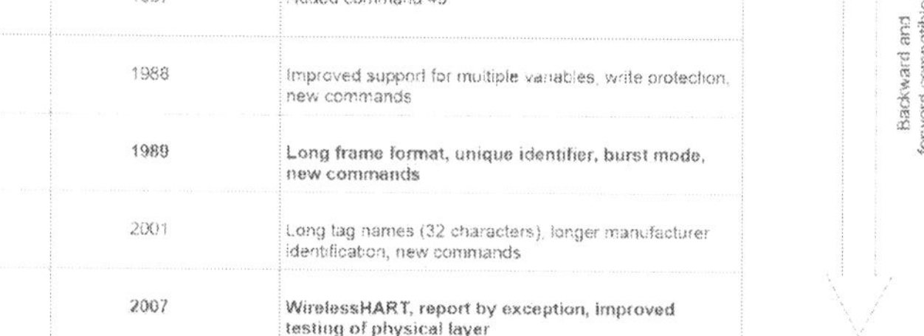

HART Versions

Revision	Year Introduced	Features
2.0	1986	First open specification
3.0	1987	Added command 49
4.0	1988	Improved support for multiple variables, write protection, new commands
5.0	**1989**	**Long frame format, unique identifier, burst mode, new commands**
6.0	2001	Long tag names (32 characters), longer manufacturer identification, new commands
7.0	**2007**	**WirelessHART, report by exception, improved testing of physical layer**

* Assuming host and slaves fully comply to the HART standard. Interoperability issues do exist

Always check your version

HART is both backward and forward compatible in theory. The assumption is that both the hosts (master) and the slaves fully comply to the HART standard.

In the real world, this not always the case. I have had pretty good luck with the backward part, but that forward part has tripped me up a few times.

This was particularly true when they started producing HART V6 and V7 instruments. Almost all the HART smart cards had issues. It then took several months for the HART smart card manufacturers to catch up.

This has left me very sensitive to versions; if I run into a problem integrating something, I check which version of the protocol the HART master is versus the version of the protocol that the slave was written to.

STATUS INFORMATION

HART has several different statuses: status, response code, communication status, device status, extended device status, condensed status (setting in the device), device variable status, device family status, additional device status.

This may appear confusing to a new user. The best way to understand statuses is in the context of HART commands. Below is a summary chart.

Status summary

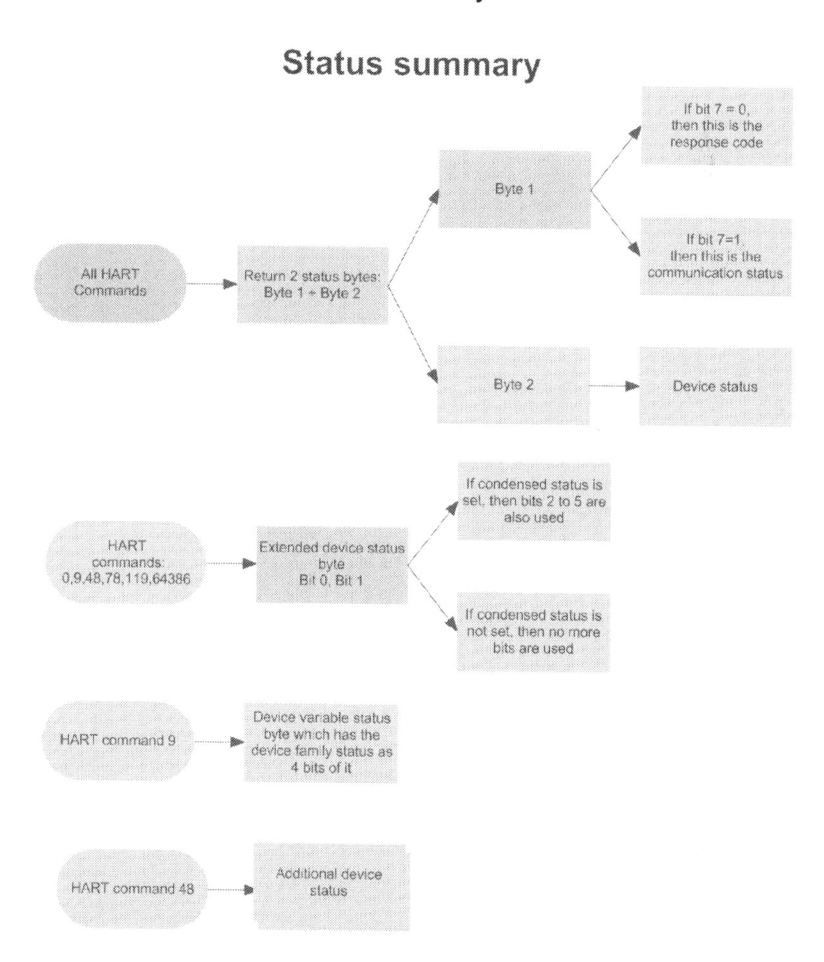

STATUS COMING BACK IN ALL COMMANDS

All HART commands return two bytes of statuses. The first byte is called the response code. If the response code is 0x00, then communications is good, and the device was able to process the command without an error. If there is a problem with communications, then the response code contains the Communication Status.

Communication Status*	
Bit Mask	**Definition**
0x80	This bit must be set to 1 to indicate a communication error
0x40	Vertical Parity Error - The parity of one or more of the bytes received by the device was not odd
0x20	Overrun Error - At least one byte of data in the receive buffer of the UART was overwritten before it was read (i.e. the slave did not process incoming byte fast enough)
0x10	Framing Error - The Stop Bit of one or more bytes received by the device was not detected by the UART (i.e. a mark or 1 was not detected when a stop bit should have occurred)
0x08	Longitudinal Parity Error - The longitudinal parity calculated by the device did not match the check byte at the end of the message
0x04	Reserved - set to zero
0x02	Buffer Overflow - The message was too long for the received buffer of the device
0x01	Reserved, set to zero

* Taken from HART Specification V7.0

If communications was okay, but there was an error in processing the command, then the response code is returned in the first byte.

Response codes*					
Value	Meaning	Value	Meaning	Value	Meaning
0	Success - no command specific error (E)	18	Invalid units code (E)	35	Delayed response dead (E)
1	Undefined (E)	19	Device variable index not allowed (E)	36	Delayed response conflict (E)
2	Invalid selection (E)	20	Invalid extended command number (E)	37-59	Reserved (E)
3	Passed parameter too large (E)	21	Invalid I/O card number (E)	60	Payload too long (E)
4	Passed parameter too small (E)	22	Invalid channel number (E)	61	No buffers available (E)
5	Too few data bytes received (E)	23	Sub-device response too long (E)	62	No alarm/event buffers available (E)
6	Device-specific command error (E)	24-27	Reserved (W)	63	Priority too low (E)
7	In write protect (E)	28	Multiple meanings (E)	64	Command not implemented (E)
8-14	Multiple meanings (W)	32	Device is busy (DR could not be initiated) (E)	65-72	Multiple meanings
16	Access restricted (E)	33	Delayed response initiated (E)	96-111	Reserved (W)
17	Invalid device variable index (E)	34	Delayed response running (E)		

* Taken from HART Specification V7.0

The second byte of the status is the Device Status. This indicates the current operating health of the field device as a whole.

Device Status*	
Bit Mask	**Definition**
0x80	Device Malfunction - The device detected a serious error or failure that compromises device operation
0x40	Configuration Changed - An operation was performed that changed the device's configuration
0x20	Cold Start - A power failure or Device Reset has occurred
0x10	More Status Available - More status information is available via Command 48, Read Additional Status Information
0x08	Loop Current Fixed - The Loop Current is being held at a fixed value and is not responding to process variations.
0x04	Loop Current Saturated - The Loop Current has reached its upper (or lower) endpoint limit and cannot increase (or decrease) any further.
0x02	Non-Primary Variable Out of Limits - A Device Variable not mapped to the PV is beyond its operating limits.
0x01	Primary Variable Out of Limits - The PV is beyond its operating limit.

* Taken from HART Specification V7.0

Before HART V6, if there were a communication error, then the device status would be meaningless. In HART V6 and above, it is now required that this value is meaningful in every response.

The charts above are for HART V7. These charts have been evolving. Bits have been added, but none removed. Some meanings have also evolved.

For example, in HART V7, a process problem like loss of echo would have the device issuing a Device Status value of 0x90 (0x80 + 0x10), meaning that you cannot trust your primary variable and that there is more status available. A HART V6 device would issue a 0x80 only if a device malfunctioned, and a loss of echo would cause only a 0x10 value.

Example:

In the following trace, the master issues a HART Command 3 to the slave device, and we see that the slave returns 00 40, which means that communications is good and that there was a configuration change. That is, the configuration change flag has been set.

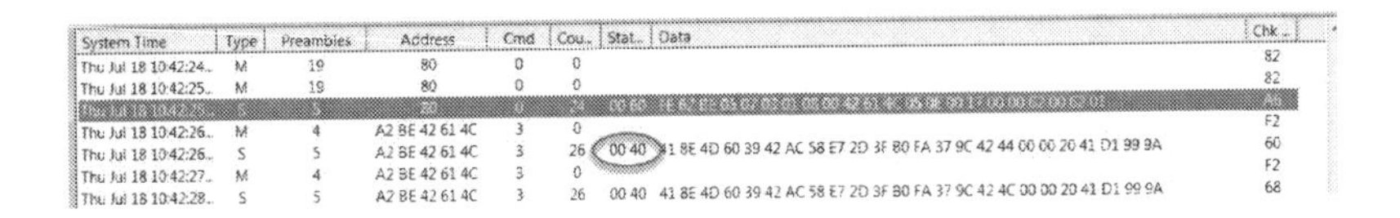

System Time	Type	Preambles	Address	Cmd	Cou..	Stat..	Data	Chk..
Thu Jul 18 10:42:24..	M	19	80	0	0			82
Thu Jul 18 10:42:25..	M	19	80	0	0			82
[illegible]	S	5	80	0	24	[illegible]	[illegible]	[illegible]
Thu Jul 18 10:42:26..	M	4	A2 BE 42 61 4C	3	0			F2
Thu Jul 18 10:42:26..	S	5	A2 BE 42 61 4C	3	26	00 40	41 8E 4D 60 39 42 AC 58 E7 2D 3F 80 FA 37 9C 42 44 00 00 20 41 D1 99 9A	60
Thu Jul 18 10:42:27..	M	4	A2 BE 42 61 4C	3	0			F2
Thu Jul 18 10:42:28..	S	5	A2 BE 42 61 4C	3	26	00 40	41 8E 4D 60 39 42 AC 58 E7 2D 3F 80 FA 37 9C 42 4C 00 00 20 41 D1 99 9A	68

EXTENDED FIELD DEVICE STATUS

This status is returned in commands 0,9,48,78,119, and 64386. This status is one byte long. Traditionally only the first two bits were used:

- Bit 0 – Maintenance Required. If this is set, the device has not malfunctioned but does require maintenance.

- Bit 1 – Device Variable Alert. This is set if any Device Variable is in Alarm or Warning state.

If the NAMUR NE107 Condensed Status selection in the field device is set, then the following five bits of the Extended Field Device status are used and mean the following:

- Bit 2 – Critical power failure

- Bit 3 – Failure

- Bit 4 – Out of specification

- Bit 5 – Functional check

NAMUR is an organization made up of several large chemical companies in Germany/Europe. As a group, they have published many different documents that they call 'Recommendations.' These are essentially specifications on how they would like something done.

NE 107 is entitled 'Self-Monitoring and Diagnosis of Field Devices.' It talks about the importance of statuses to the operation of a plant and how best it should be done. When NAMUR NE107 Condensed Status is turned on in a HART device, the additional bits in the Extended Field Device Status results in the device complying with the recommendations in NE 107.

Example: Command 0: read unique identifier

This command returns the following data:

- Byte 0: 'FE' (expansion)
- Byte 1-2: Expanded device type
- Byte 3: Minimum number of preambles required
- Byte 4: Universal command revision
- Byte 5: Device-specific command revision
- Byte 6: Software revision
- Byte 7: Hardware revision
- Byte 8: Device function flags
- Byte 9-11: Device ID number
- Byte 13: Maximum number of device variables
- Byte 14-15: Configuration change counter
- Byte 16: Extended field device status
- Byte 17-18: Manufacturer identification code
- Byte 19-29 Private label distributor code
- Byte 21: Device profile

The returned message is shown below, where the extended field device status is 0:

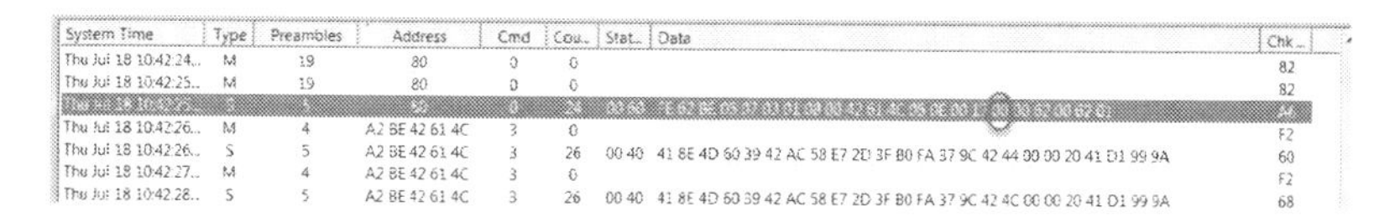

System Time	Type	Preambles	Address	Cmd	Cou..	Stat..	Data	Chk..
Thu Jul 18 10:42:24...	M	19	80	0	0			82
Thu Jul 18 10:42:25...	M	19	80	0	0			82
[illegible]	[illegible]	5	80	0	24	[illegible]	[illegible]	[illegible]
Thu Jul 18 10:42:26...	M	4	A2 BE 42 61 4C	3	0			F2
Thu Jul 18 10:42:26...	S	5	A2 BE 42 61 4C	3	26	00 40	41 8E 4D 60 39 42 AC 58 E7 2D 3F B0 FA 37 9C 42 44 00 00 20 41 D1 99 9A	60
Thu Jul 18 10:42:27...	M	4	A2 BE 42 61 4C	3	0			F2
Thu Jul 18 10:42:28...	S	5	A2 BE 42 61 4C	3	26	00 40	41 8E 4D 60 39 42 AC 58 E7 2D 3F B0 FA 37 9C 42 4C 00 00 20 41 D1 99 9A	68

Example: Command 9: Read up to eight device variables with status

HART command 9 was added in HART V6 as a universal command. It is used to read any variable in a device. It is typically used to read dynamic variables.

This command returns the following data:

- Byte 0: Extended field Device status
- Byte 1: Slot 0: Device Variable Code
- Byte 2: Slot 0: Device Variable Classification
- Byte 3: Slot 0: Units Code
- Byte 4-7: Slot 0: Device Variable Value (float)
- Byte 8: Slot 0: Device Variable Status
- Byte 9: Slot 1 begins (bytes 1 to 8 repeated for slot 1)

In this command, there are two different types of status returned: the Extended Field Device Status, which is described above, and the Device Variable status, which is described below.

DEVICE VARIABLE STATUS

Device Variable Status is a measure of the overall health of the variable being read. The chart below gives the general meanings of this status:

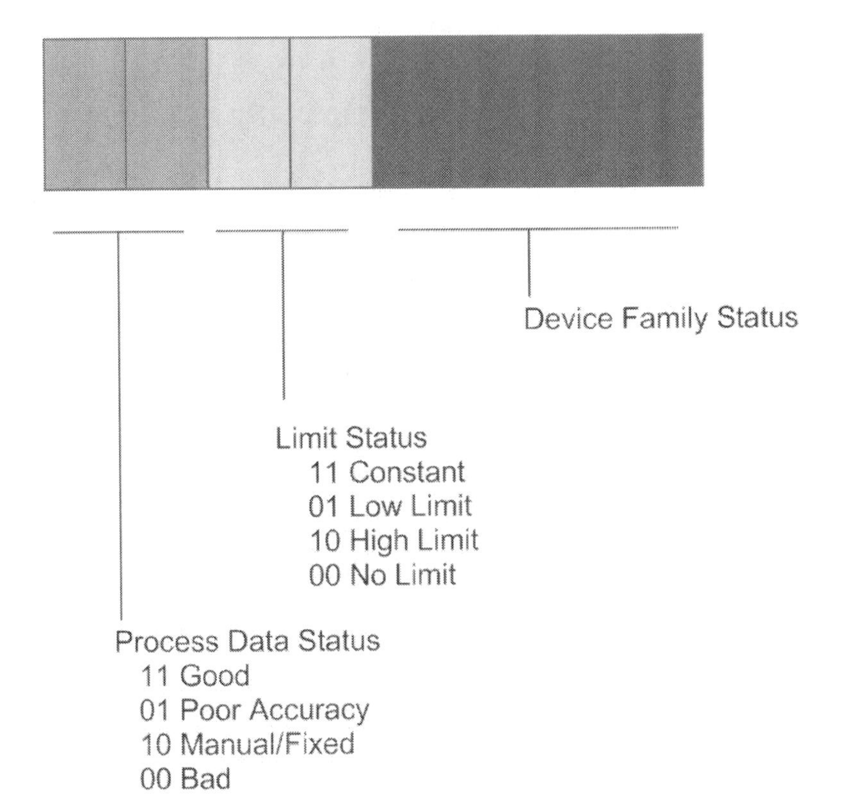

DEVICE FAMILY STATUS

Device Family Status is part of the Device Variable Status.

HART has a series of 'Device Family Specifications' for different types of field devices such as temperature, level, flow, etc. Each 'family' will define the meaning of these four bits and may define the entire byte. These specifications have been at various levels of release, so many vendors have defined their codes here in the absence of a released specification.

For decoding the Device Variable Status and the Device Family Status, it is best to consult either the manual or the HART field device specification document (Lit 18) for the device.

ADDITIONAL DEVICE STATUS

HART command 48 is used to read Additional Device Status. The response message contains 25 bytes of data. Bytes 0 to 5 and 14 to 24 hold the Device-Specific Status which corresponds to the Error Display Codes shown on the screen of the device. The mapping is related to bit location of the bit value '1' in the Device-Specific Status.

For example, if there is a '1' located in bit 4 of byte 0 of the response message, then that will correspond to an error code 4. If byte 2 of the response message has a 1 in bit 0, then it would be code 2*8+0=16.

HART command 48 has also evolved. It was a Common Practice Command, and now in HART V7, it is a Universal Command. It has also expanded in size as well, adding more bits to extend the possible error messages.

To properly decode this, you will need either the device manual or the HART field device specification document (Lit 18) for the device.

Choose your system well

If you read through the status information section and feel a bit confused, then welcome to the club.

The status information was grown over time, and as a result, there are multiple ways of reporting errors.

This means that if the design of both the instrument and host is not done with some forethought, then you can end up with a 'flood' of alarms when something goes wrong. That is, you will have several alarms all saying the same thing.

This is where alarm management comes in, and it is no small task. However, it is worth doing.

HART smart cards offer different access levels to this information. Some will provide this information in a table. Others will require that you issue commands to the card. This makes choosing which card to use important.

Please don't think that I am just picking on HART with this comment—all the older protocols have alarm management issues. The newer ones like PROFINET have learnt from the past and have this alarm management in much better shape—it too can benefit from alarm management, but it is an easier task.

Conclusion

Statuses are key to getting the 'value' out of HART. Unfortunately, this chapter is not the easiest to read—it was certainly not the easiest to write.

As you implement good alarm management in your plant, you may want to re-read this section once or twice to see what information is important to you. Related to alarm management is how your plant manages its assets, which is the next part of our journey.

CHAPTER 3: ASSET MANAGEMENT & CONFIGURATION SOFTWARE

"What's measured improves."
--Peter F. Drucker

What is measured, improves, is generally true—the opposite, however, is certainly true: what is not measured or looked at does not improve.

Asset management is like that. It needs to be looked at. If you want to get the most out of your plant, then asset management is key—and configuration software is one of your key tools.

Asset management is a big topic that has many definitions and covers many aspects of plant operations. Whole books could and have been written on it. The same goes with configuration software.

In this chapter we will give you a brief look at some of the key aspects of asset management and configuration software, focusing on what is pertinent to HART as well as practical.

ASSET MANAGEMENT

Traditional asset management has divided asset management into three elements:

- **Human assets**: the employees in a company and the knowledge base that they maintain.
- **Virtual assets**: the processes and information in place to manage the physical assets and human assets.
- **Physical assets**: the actual devices and objects in the plant.

(From: Snitkin, Sid. Asset Lifecycle Management – A New Perspective on the Challenges and Opportunities, ARC White Paper, July 2008.)

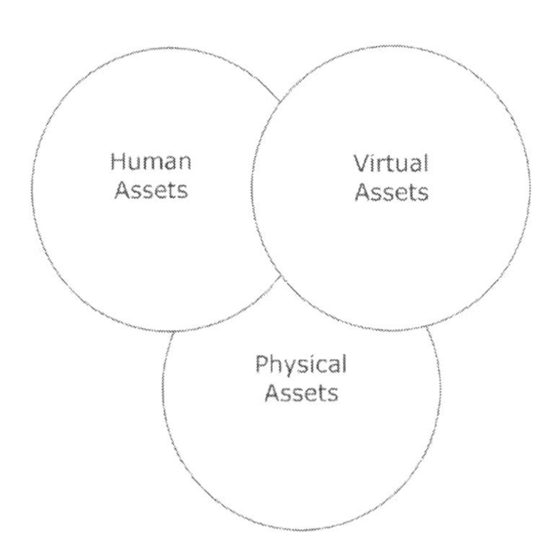

At its heart is the goal of getting the most out of the investments in a plant. As with most optimization plans, knowledge is critical:

- Can I trust my process variable?
- Is my instrument healthy?
- Is the process running within limits?
- Am I getting all the information?
- If there is a problem with an instrument, how can I fix it, quickly?

This definition is extending the traditional definition of asset management to include and focus on data.

The fact that HART provides two-way communications between your instrument and control system (when installed with a HART Smart Card) means that it can help answer all the above questions.

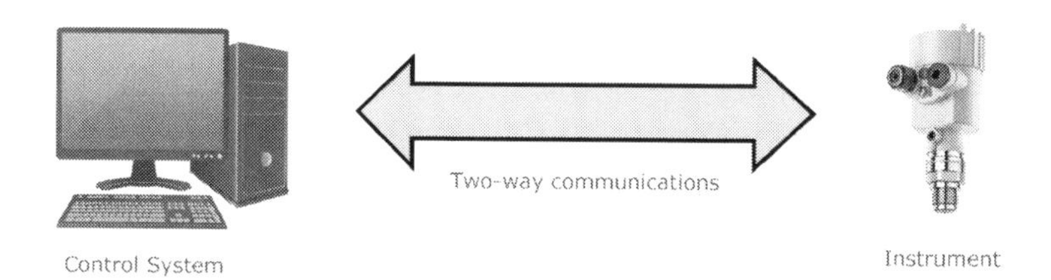

The status bytes which we discussed in the previous chapter provide information on the health of the process variable and the instruments—and in some cases, the process itself.

The status bytes which we discussed in the previous chapter provide information on the health of the process variable and the instruments—and in some cases, the process itself.

Therefore, monitoring this is key. One issue with HART is that unlike many other protocols, this information is not always read and monitored. Quite often, you have to design your system to do this. The value of doing this should be obvious, but this does need to be done. Again, this is where the HART Smart Card comes in. In addition, the system will need to read and display this information appropriately.

Some instruments have more accuracy than what can be transmitted via 4-20 mA. However, using the HART digital channel to read the process variables will provide you with all the accuracy of the instrument.

When you are monitoring your instruments and you notice that an instrument is having a problem, how do you quickly resolve it?This is where configuration and diagnostic software and handhelds come in. This software communicates with your instrument and tells you what is wrong, and in its database will be key information for getting technical support or replacement of the instrument. Information such as:

- Serial number
- Model number
- Manufacturer
- Software revision
- Hardware revision

◆ ◆ ◆

But … there is a catch!

Asset management sounds good. The costs benefits are huge, but there is a catch: you have to design it into your business, from the instrument up to the CEO.

This is not easy.

The HART protocol, for example, meets all the requirements for good asset management.

However, you need to have HART wired into HART smart modules. You need to have good HART configuration software in place. Your people need to know how to use the software and the instruments. They have to know enough about the HART protocol to communicate with all of your instruments. Preventative maintenance has to be the norm.

The HART website has site examples from *HART Plant of the Year Awards*. When you read through the success stories, you see that all of these plants did pretty much all of the above.

CONFIGURATION/DIAGNOSTIC SOFTWARE AND HANDHELDS

There are many software packages and handhelds on the market that can be used to configure and troubleshoot a HART instrument. The three big software packages on the market are Emerson's AMS, Siemens PDM, and PACTware. However, there are many more.

The handheld market used to be dominated by Emerson's 375/475 HART Communicator, which has now been replaced by AMS Trex^{TM}. However, now there are many competitors: Meriam, ProComSol, ABB's DHH805, to name a few.

In general, the end users have many options open to them.

These software and handhelds use either DD (Device Descriptions) or DTM (Device Type Manager) to be able to communicate to a HART instrument. When setting up one of these devices or packages, the user will have the choice between either a generic DD/DTM or a device-specific DD/DTM.

Please note that Device Descriptions (DD) are also sometimes called Electronic Device Descriptions (EDD). The HART standard refers to them as DD, but the international standard that they are now based on refers to them as EDD.

If the generic DD/DTM is picked, then the software only has access to the universal and common practice commands. This means two things:

- The user may be limited in what they can do. This depends on the device manufacturer. Some have implemented many—if not all—of the Common Practice Commands, while others have not implemented any.
- The user may see some communications errors if they use a common practice command that is not supported by the instrument they are connected to.

A device-specific DD/DTM will have access to all the commands but will mainly use the device-specific commands. These commands are related to the firmware revision of the device.

As the device is developed over time, sometimes new device-specific commands are added to accommodate new features. Therefore, the device-specific DD/DTM is linked to the correct firmware version of the instrument. However, HART also has a backward and forward compatibility concept built into the standard.

This means that in theory, the end user should be able to use newer DTM/DD on older devices and older DD/DTM on newer devices. In practice, this is not always the case.

A DD contains information on:

- All parameters in an instrument
- All parameter interactions
- How to read and write those parameters to/from the instrument
- Simple procedure for setup and diagnostics

DTMs contain the same information as the DD. The only difference is that a DTM is a program while a DD is a series of text files. Which one you use depends on what you are using.

For example, AMS Trex Device Communicator™, AMS and SIMATIC PDM uses DDs, while PACTware and Fieldcare uses DTMs. To make things a little bit more confusing, AMS can use DTMs and there is software you can get for PACTware that will let you use DDs in some cases.

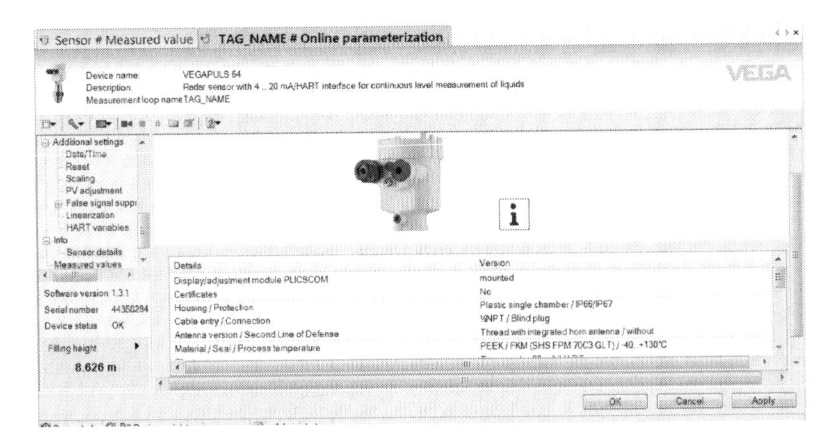

A third alternative newly released is called Field Device Integration (FDI). FDI is the marriage of DD and DTM technology, presented as their natural evolution them. The big step forward is that there is a common interpreter to all software packages and a common file format.

Therefore, one FDI package will be able to be loaded into many different software packages. For example, the same FDI package for a radar device could be loaded into both SIMATIC PDM and Emerson's AMS. This is certainly a step forward.

FDI is hoped to be an evolution and not a third standard. Both SIMATIC PDM and Emerson's AMS are migrating to FDI while maintaining their DD support for backward compatibility.

DDs and DTMs, oh my!

If you think this sounds a little messy, then you are correct!

We currently have two standards out there—DD and DTM—with a third one just released.

They all do the same job and have to be linked with the firmware version of the instrument.

Also, both DD and DTM standards have been evolving, so these things are also linked to the version of configuration software you are using.

On the other hand, when this technology works, it is a thing of beauty. You have access to a lot more information, and troubleshooting instrument issues are easier and faster.

I used to work for a company who made level transmitters. Climbing a 40-foot tank and leaning over the top to use an infrared handheld to look at parameters in the device is not a lot of fun. This is especially true if you add in rain or snow!

However, sitting in a control room and being able to configure the instrument over the network is fine in any weather.

PARAMETER INTERACTIONS

Many devices will have one parameter that will cause changes to other parameters. For example, setting an application parameter from solids to liquid may cause the speed of response to be changed. These interactions are programmed into the device and the DD/DTM. They should match. However, sometimes they don't.

Also given these dependencies, the order in which you write down the parameters can have an impact on what happens. For example, if parameter J changes parameters V and Z, then the programming of the DD/DTM is different than the device. The results of downloading the parameters from A to Z may be different than if you download Z to A.

This is a problem for device vendors. If the device vendor has done their job correctly, then the end user does not need to know about this. However, in the real world, it is useful to keep this in mind and verify that all the parameters are what you think they should be.

This is one of the reasons that it is a very good idea to read all the parameters from a device after a write—to verify that they were written correctly and that no strange parameter interaction has caused an unexpected result.

ONLINE/OFFLINE

The different configuration software has all approached the topic of online/offline differently and is one area where end users can run into difficulty.

It is very important to know if the screen you are on is online or offline—and if you are offline, whether the parameters match what is in the device.

AMS is all online, and for the most part, you change one parameter at a time. All your offline data is stored in a database, which you can retrieve based on date. This is perhaps the most straightforward method from an end user's point of view.

PDM is mostly offline, and you have to read and write the parameters to the device. There are icons to show if the online parameters match the offline parameters. There are some online dialogue boxes. Typically, it is pretty easy to tell which screen you are on, but you do need to think about it.

PACTware is a mix of offline and online dialogue boxes. This is the one program where you have to be very careful keeping track of what kind of screen you are on. If there is an icon on the screen that looks like a circle with a arrows, then that PACTware screen is online. Otherwise, the screen you are on is probably offline.

In PACTware it is very common for beginners to bring up the main parameter listing, make a change, and wonder why the instrument has not changed. Again, typically, it is pretty easy to tell what kind of screen you are on. However, you have to be aware ahead of time that there are two types of screens.

REVISIONS

DD/DTMs are linked to the firmware version of the device and sometimes with the version and type of software. For example, in the description of the DD, it may say that it is for device Firmware 3.5 and for PDM version 7.0. This means that it was tested in PDM version 7.0. It may or may not work in PDM Version 6.0 or PDM Version 9.

The only way to really know is to look through the manufacturer's download page and see if there is another DD for that firmware version for the later version of PDM. Ideally, you have the correct DD/DTM for both the version of the software you are using and the firmware your device has. This is the best case situation.

However, even if this is not the case, you may still have some, if not all, functionality by using either a DD/DTM for a different version of firmware or by using a generic DD/DTM. Always keep in mind HART's backward and forward compatibility feature and see what happens. Even though HART has this feature, sometimes the software package you are using will not let you try a different version.

Please note that using anything other than the right DD/DTM for the right firmware for the right software has some risks. There is a chance of messing up the device. However, these risks are very low, and a factory reset gets you out of most issues.

SOFTWARE COMMON FEATURES

Configuration and diagnostic software and handhelds used to vary a fair bit in functionality. They still do to some extent. However, most of them now have the following list of features:

- Support for quickstart wizard
- Visualizing the main process variable (including graphing)
- Full identification of the device (serial number, model number, tag name, etc.)
- Advanced diagnostic (echo curve or valve signature curve)
- Calibration wizards
- Offline storage of parameters
- Method for comparing parameters
- Advanced networking (ability to connect through other networks)

Quickstart wizards are designed to easily set up the device for between 80 to 90 percent of applications. It will ask you simple questions, normally with drawings, to show you what it wants. These are designed to minimize setup time and reduce setup errors.

Visualizing the main process variable and other variables lets you see the value remotely. The graphing function is very useful in troubleshooting process issues by giving you a cheap chart recorder.

This function cannot run for a long time—however, if you want to track the variable over an hour or two, then this function works well.

Each HART device has identification information in parameters that the software can read and display for you. When troubleshooting the device, this is very important.

Advanced diagnostics are critical for some devices and applications. The two prime examples would be the valve signature curve for valve positioners and echo profiles for radar and ultrasonic level devices.

In the case of valve positioners, the end users can use the valve signature curve to determine when they need to replace a valve. Making use of this feature can save some sites a lot of money due to the high cost of valves.

For radar and ultrasonic level devices, one of your main troubleshooting diagnostics is viewing the echo profile. In fact, for some applications, having access to the echo profile is the only way to adjust the device to work correctly.

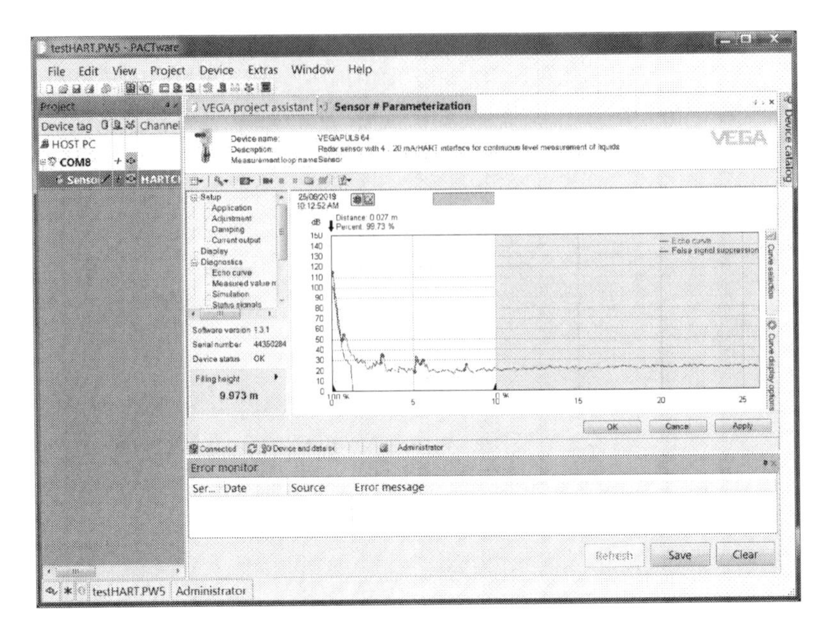

Calibration wizards are a requirement in some industries. They work by taking you through a process of verifying that the value coming through the 4-20 mA channel matches what is in the instrument and that the range is correct. There is normally also a trim function where you can adjust the value to overcome drift.

Offline storage of parameters is required for quick device replacement. If you have to program a new device, you will need to know what the values were in the previous device. Offline storage gives you that information.

Comparing parameters is important for troubleshooting. Field service uses this feature a lot. They go to a site and get a device set up and working fine. Then they save the parameters, and if they are called back because the device is not working, they use the compare function to see what has changed.

Advanced networking is something that all the configuration software and even some of the handhelds have. This is referring to the ability of the software to communicate through different communication protocols to get right down to the instrument level.

For example, in the drawing on the following page, the maintenance computer is connected to HART instruments via Industrial Ethernet and a HART Gateway. This is all possible with AMS, PDM, and PACTware.

However, the gateways that support this are different. That is, a Siemens gateway will let PDM drill through, but not AMS or PACTware—while many smaller vendors' gateways will support PACTware but may not support PDM or AMS.

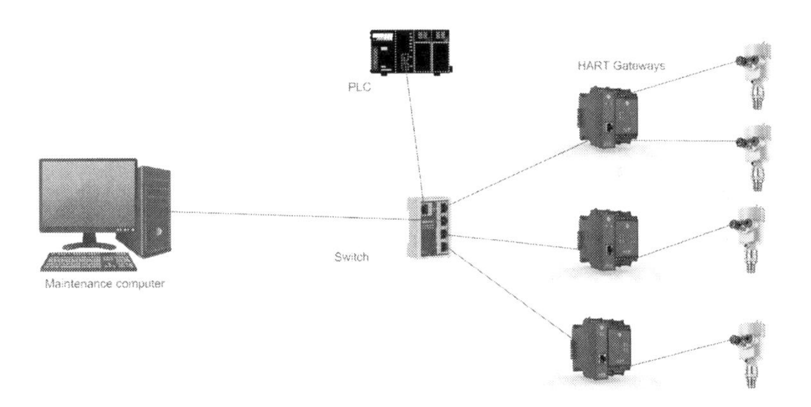

In getting through the gateway, the vendors can choose different methods. HART has HART-IP, which is vendor-neutral. This means that some gateways—for example, Thorsis Technologies isNet lite, which supports HART-IP—can support any software that uses HART-IP. HART-IP is supported by both AMS and PDM. It also has a communication DTM for PACTware support.

Conclusion and next steps

This has been an introduction to asset management and configuration software. Asset management can and should be studied in greater length. Configuration software is something that needs to be tried out and explored with your technical staff. What works for one company may not work for another due to different organizational culture, types of devices, or networking.

Next up are three chapters that focus in on the three most popular physical layers: HART over copper, HART-IP, and WirelessHART. In these chapters we look at the practical aspects of using these.

CHAPTER 4: HART OVER COPPER (FSK PHYSICAL LAYER)

"Did I do anything wrong today," he said. "Or has the world always been like this, and I've been too wrapped up in myself to notice?"

--Douglas Adams

HART over copper is everywhere. Like many things that are everywhere, it is easy not to notice it. HART instruments are purchased as 4-20 mA instruments, and the HART aspect may or may not ever be used.

This is like buried treasure in your process plant. Learn to connect to it, and you can start benefiting from all the information it can provide. In this chapter we will go into the details you need to know to connect HART over copper.

HART OVER COPPER

When someone says they have a HART instrument, they are typically talking about HART over copper. The HART digital signal rides on top of the 4-20 mA analog channel.

If HART is used by itself, then the analog signal is set to one fixed milliamp value. HART uses Bell 202 frequency shift keying (FSK) to superimpose the message on top of the analog signal.

The digital signal uses two different frequency bursts to represent a binary '1' and '0'. A '1' is encoded using a 1200 Hz frequency burst, and a '0' is encoded using a 2200 Hz frequency burst. The mean value of the burst is zero, so the reading of the slow-moving 4-20 mA signal is generally not affected.

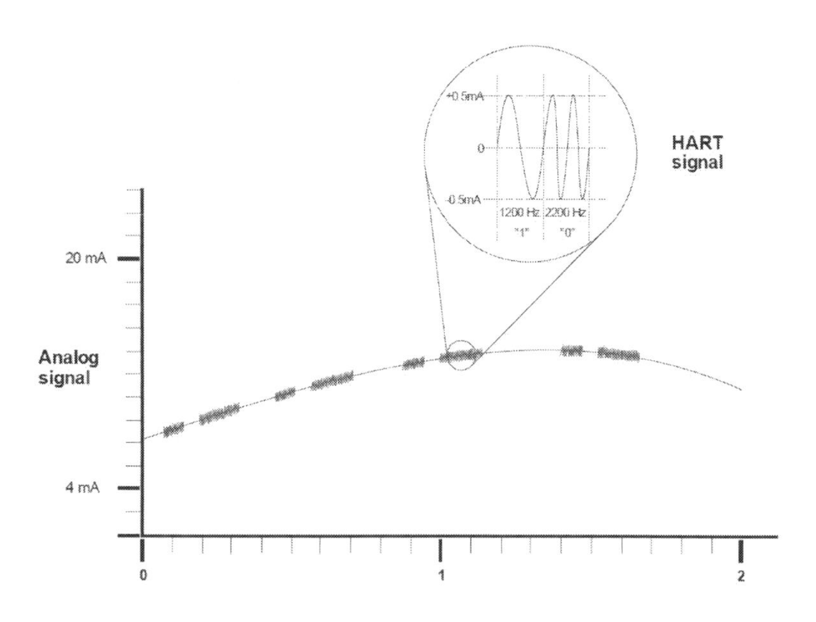

This signal is filtered out of most analog input cards. However, although rare, incompatibilities do exist.

Using an oscilloscope, you can see the HART signal by clamping leads across the HART resistor (approximately 250 ohms) and then setting it to 100 ms/div and zooming into the signal. The actual voltage level will depend on the output of the transmitter and the size of your HART resistor.

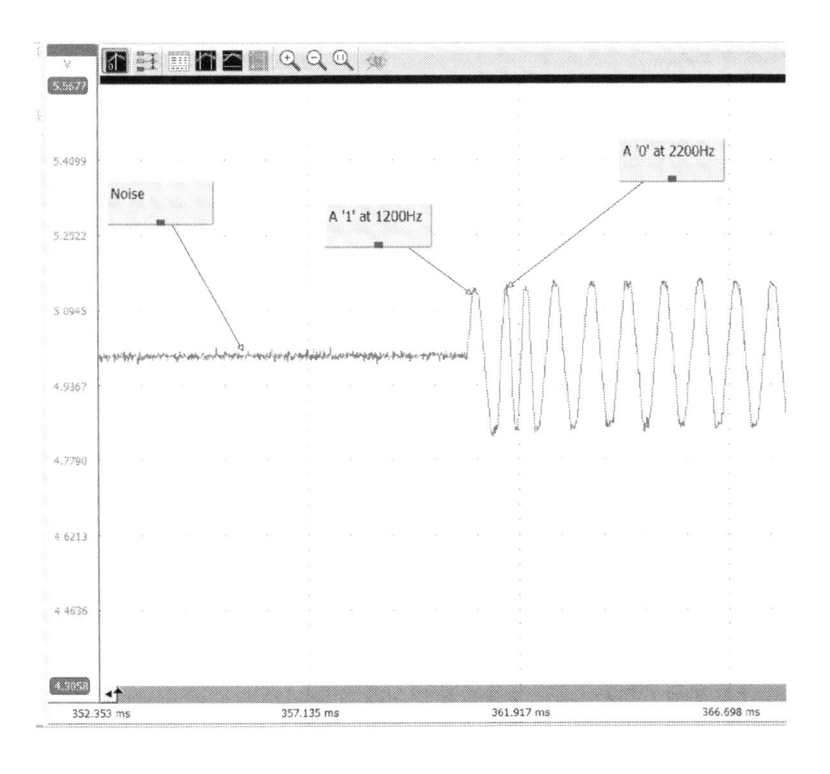

The elephant in the control room

4-20 mA has been around for so long that I compare it to the elephant in the control room: we as engineers never really look at it anymore.

However, when you start to examine the accuracy of it (which I did in an article published on profibus.com), then it is clear that most instruments have more accuracy than can be transmitted by 4-20 mA.

Digital communication protocols like HART, PROFIBUS, and Foundation Fieldbus can transmit the full accuracy of your instruments.

BASIC DESIGN RULES

HART design rules are not widely known. Many believe it is the same as 4-20 mA and does not require any extra considerations. However, although the rules are simple, they do exist and need to be considered.

There are five key things to be concerned about:

1. You need a minimum loop resistance.
2. Do not exceed the maximum Resistive-Capacitance (RC) value for the circuit.
3. All instruments must have a minimum operating voltage.
4. The cable matters.
5. The power supply matters.

Minimum loop resistance

For the HART signal to form, there has to be a minimum loop resistance of 250 ohms. Many HART input/output cards will have this resistance built in.

However, if you are using a HART modem, you need to know how long the wire is and the resistance of the wire so that you know if you need to add any resistance. For short runs under 100 metres, it is safe to assume that you need to add a 250-ohm resistor.

Maximum RC value

This rule is not widely known. For HART communications to occur, you must have an RC value of the loop, less than 65 µs. The RC value is the total resistance of the circuit multiplied by the total capacitance. The first rule sets the minimum resistance, and this rule sets the maximum. In both cases, the more complex the network, the harder this calculation is, and the more important it is. Again, if it is a point-to-point topology and the run is less than 100 metres, this is not a concern.

Minimum operating voltage

For a HART instrument to power up, it needs a minimum amount of voltage. The same applies for PROFIBUS PA and Foundation Fieldbus instruments. However, unlike those newer protocols, HART does not have a standard minimum voltage for all instruments: every instrument may have a different value. Therefore, you have to be aware of what this value is for all the instruments that you are using.

Also, you must make sure that this minimum voltage is exceeded throughout the current range. Otherwise, the instrument may power up at 4 mA fine, but then when the PV changes and the output value goes up near 20 mA, the increased voltage drop on the line will cause the voltage at the instrument to fall below the minimum voltage, causing the instrument to shut down. This scenario, unfortunately, happens more than it should.

Like the RC rule, this rule also sets a value for a maximum loop resistance. This is typically not an issue when the runs are less than 100 metres and you only have one instrument involved.

Cable

Too often, I have heard that the cabling is the easy part of the network. This is never true: the choice of cable matters.

As the first two rules imply, the type of cable used will impact the loop resistance and the RC value. It is true that for very short runs, HART will typically run on almost anything. However, put that short run near a Variable Frequency Drive (VFD), and you will experience problems that would disappear with the correct cable.

As distance increases, the importance of your choice of wire increases. The better the cable, the better the noise immunity. In general, you want to pick a shielded, twisted pair instrument-grade wire and perform the calculations before you buy and install the wires. It is also a good idea to buy a cable designed for HART (Belden 3105A is an example of such a cable).

Power supply

Many also believe that any power supply will work. This is generally true because most industrial power supplies do meet the HART power supply requirements.

However, generally true is not the same as true all the time. Therefore, it is a good idea to check the power supply that you are about to use to make sure it has the following requirements:

1. Maximum ripple (47 to 125 Hz) = 0.2V p-p
2. Maximum noise (500 Hz to 10 kHz) = 1.2 mV RMS
3. Maximum series impedance (500 Hz to 10 kHz) = 10 ohm

Please note that if you get a power supply that does not meet these requirements, the HART instrument may still work. However, it probably will not work all the time. In the past, this has not been the easiest type of technical support issues to resolve.

Instrumental differences

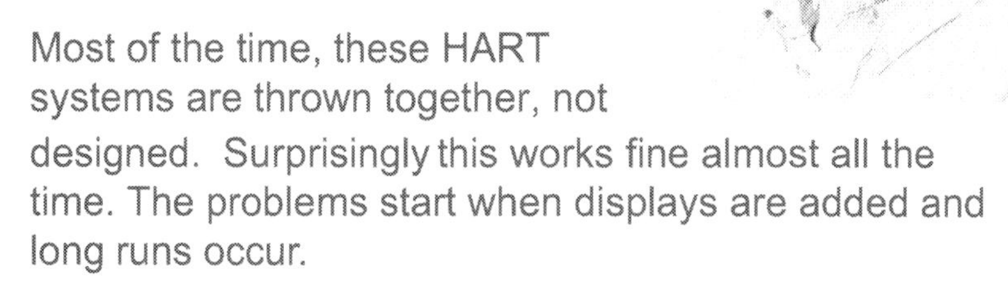

Everyone who has worked with HART knows about the minimum loop resistance. However, the fact that you can have too much resistance causing the RC value to go too high was a big revelation for me.

Most of the time, these HART systems are thrown together, not designed. Surprisingly this works fine almost all the time. The problems start when displays are added and long runs occur.

Minimum operating voltage was another thing that most people do not think about until a device does not power up—or worse, it shuts off when the value goes high (which, by the way, was how I first ran into this issue).

My big surprise with minimum operating voltage was that there was no standard for this in HART. All instruments are different.

HAZARDOUS ENVIRONMENTS: INTRINSICALLY SAFE, EXPLOSION-PROOF

A hazardous environment is one where a spark can cause an explosion. There are two popular methods for handling hazardous environments: explosion-proof and intrinsically safe.

The explosion-proof method is a containment method where the wires are all encased in metal piping that is sealed at both ends—so that if there is an explosion, it will be 'contained' in the piping/conduit. This method works, but is very expensive to install and is subject to human error—the sealant is very hard to work with so it is not uncommon for an installer to 'forget' to install it in all the locations.

Intrinsically safe is a method where the energy sent to the field is limited so that there is not enough power to cause a spark. This method uses barriers to limit the energy going to the field device. The barrier is installed in the 'safe' area. This is the preferred method.

There are a number of different barriers available. Make sure you use one that has been designed for use with HART protocol, as some of the standard 4-20 mA barriers will kill the HART signal.

Grounding

Most installations you see use traditional grounding methods for 4-20 mA loops and HART, where the cable shield is only connected to ground at one end. The reason for this is to provide noise a path to ground—and to prevent ground loops.

However, international standards recommend grounding the shield *at each end* for point-to-point and *at every point* for multi-drop (described in IEC 61000-5-2 Electromagnetic compatibility – Part 5, Installation and mitigation guidelines – Section 2: Earthing and cabling). See figure below for details.

This use of an equipotential grounding grid and grounding the shield at every point provides a low impedance to ground at the frequency of concern as detailed in IEC 50310. Simply put, this is a better way than the 'traditional' method.

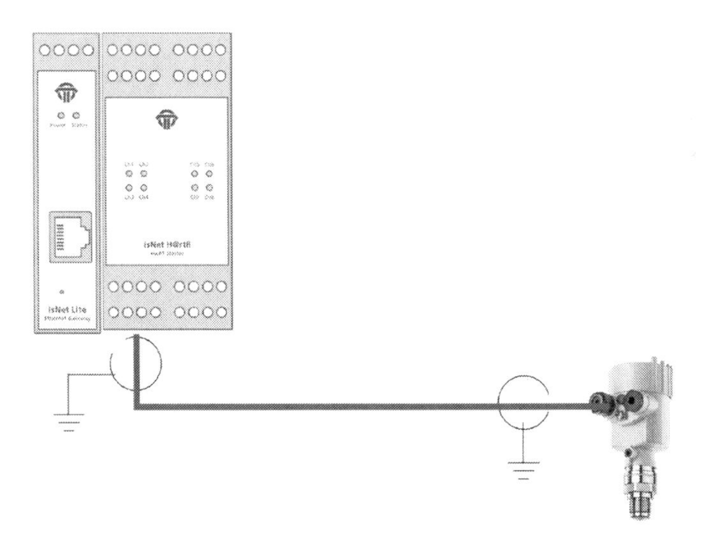

When you're right, you're right

I was once told that grounding methods was like religion: never tell them that they are wrong.

However, in this section, I have effectively told most North Americans that their favourite grounding method is wrong!

Sorry, but if you read the IEC document that I reference above, you will see that I am right. However, if you use a good quality shield instrument cable and just ground at one end, things should still be fine.

Having an equal potential grounding grid and grounding at both ends is better, but you've got to do what you've got to do.

Cable separation

Best practice for installations worldwide typically states that you should separate your cabling into low voltage, medium voltage, and high voltage runs.

Also, that there is separation between these runs either through distance or via a grounded cable tray. Below are the requirements from the Ontario Electrical Safety Code.

Where you lay your HART cable has a big impact on the amount of noise induced into that cable. Communication cables and low voltage cables (24V DC) can be put into the same cable tray without an issue. However, if they are near a medium or high voltage line, then current will be induced and bit errors will occur.

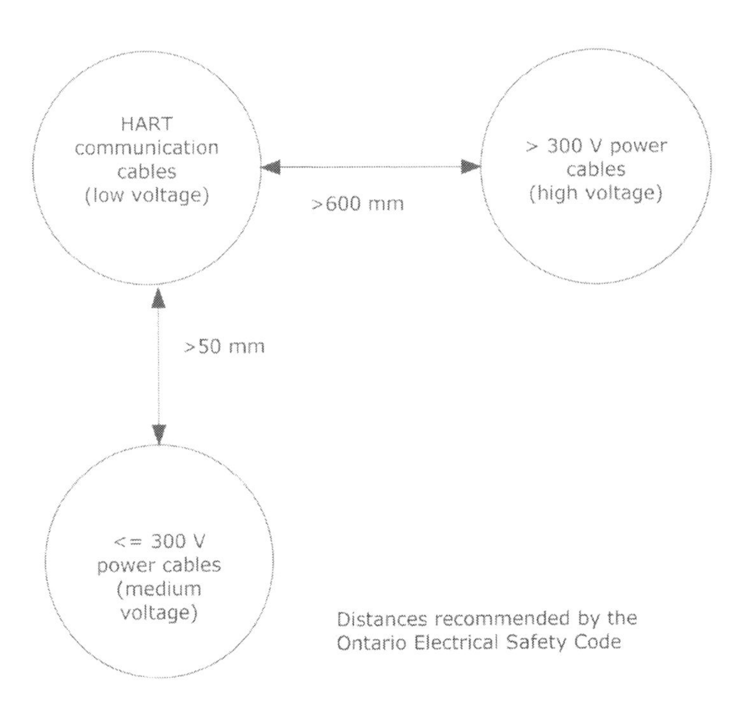

Keep 'em separated!

In almost all site audits that I've done, I have found one or many areas where not only has the cable separation rule been violated, but more often than not, the communication cable has been tie-wrapped to the power cable.

In almost all site audits that I've done, I have found one or many areas where not only has the cable separation rule been violated, but more often than not, the communication cable has been tie-wrapped to the power cable.

Yes, it looks nice, but induction is real and will cause you communication issues. This is particularly the case with HART. Looking at the HART signal shown at the beginning of this chapter, it is pretty easy to see that it would not take much noise to cause problems with that signal.

Update rates

There is a perception that the 4-20 mA channel for HART is updated instantly. Unfortunately, this is not true.

There are three delays:

- The first is in the instrument where there will be a small delay caused by the digital-to-analog conversion (DAC). This value is typically not published separately and does vary from instrument to instrument based on the DAC used. Generally, this is a small delay.
- The next delay is in the transmission itself. This delay is extremely small and is where the perception that 4-20 mA is instantaneous likely comes from.
- The third delay is the delay at the analog input card where the analog signal is converted into a digital value. This is a published value and varies greatly between different types of input cards and manufacturers.

This update rate should be considered when choosing input cards. Practically because this is where most of the delay is.

Also, the speed of the card is related to the cost of it. Generally, the more you pay, the faster the processing speed.

The HART digital channel is slow. It can handle between two to three messages per second. Basically, you are looking at a 0.5 second update rate for most instrument when wired point-to-point.

Obviously, the more instruments you need to talk to, the slower the overall update rate will be. Therefore, when multi-dropping instruments, the number of instruments will have a large impact on the update rate. This rate should also be considered if you are using a multipoint HART input card where the HART channel is multiplexed between the different points.

Point-to-point topology

Point-to-point is the topology where HART is at its best and is the most common topology. If you are going point-to-point, using shielded HART wires, have good grounding, and keep your runs under 100 metres—then HART will work very well.

In the Basic Design Rules section, we outlined what you need to consider:

- Minimum loop resistance
- Maximum RC value
- Minimum operating voltage

For runs under 100 metres, all you have to do is use a 250-ohm resistor and a 24-volt power supply (plus shielded wires and following good grounding practices), and you are good to go.

The RC value and minimum operating voltage are only a concern if you have a long run or if you are putting secondary devices, such as 4-20 mA display, in line as well.

A 2-wire device draws power from the wire (loop powered), whereas typically a 4-wire device will provide power to the wire (active source). Some 4-wire devices can be configured to be either loop-powered or active source.

The wiring is different if you have a loop powered (passive) device or an active source:

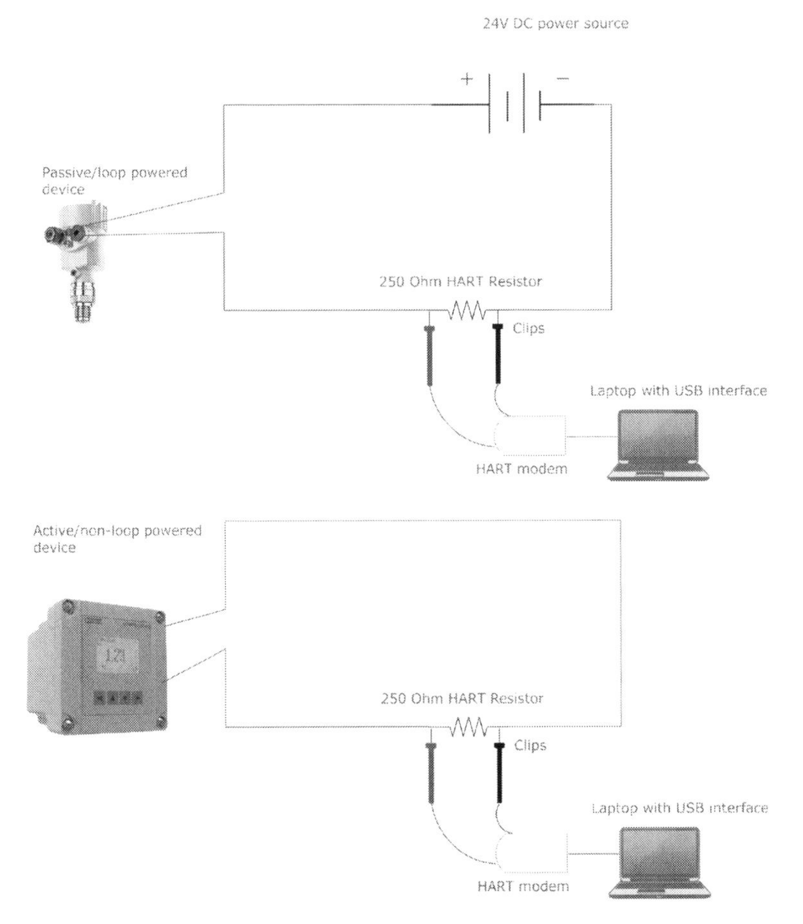

Going the distance

Point-to-point is where HART shines. Keep the distance from the HART Smart Card to the instrument down to a reasonable distance and HART is bulletproof. However, long runs can be problematic.

The HART specification says that HART can go 1000 metres—yeah, right. In theory, yes, but in practice, that is hard.

50 metres? Definitely. 100 metres? Sure. 200 metres? Probably. But as the distance increases, so does the probability of failure!

Multi-drop topology

Multi-drop topologies are not as commonplace as point-to-point, but they have been gaining some popularity with connecting HART instruments to a WirelessHART adapter. Customers are choosing this method for the cost savings.

However, there are two issues with multi-dropping HART:

- The update rate is very slow.
- You have to design the network.

As is mentioned in the Update Rates section, the communication rate for HART messages on the digital channel is about two per second. When HART devices are multi-dropped, only the digital channel can be used for the Process Variable.

When you multi-drop two devices, your update rate will be once per second; four devices will take two seconds, and so on. As the number increases, your update rate increases.

High update rates are not acceptable in many applications but may be in some non-time-critical applications. The non-time-critical applications are where HART multi-drop can be used.

The HART addressing method permits devices to have an address of between 0 and 63 in HART V6 and above. Earlier revisions of HART (V5 and before) only permitted addresses from 0 to 15. However, for all practical purposes, it is hard to multi-drop much more than eight devices due to design issues and current loads.

When designing a HART multi-drop network, you have to know:

- Electrical characteristics of the wire being used
- Startup current of all devices
- Steady-state current of all devices
- Minimum startup voltage of all devices
- Power source of devices used (loop powered or active source)

NOTE: When using configuration/diagnostic software such as PACTware or SIMATIC PDM on a HART multi-drop network, only one instrument can be viewed at any one time.

The wiring will differ greatly depending on whether all devices are loop powered, active source, or a combination of the two. See connection diagrams for each on the following page.

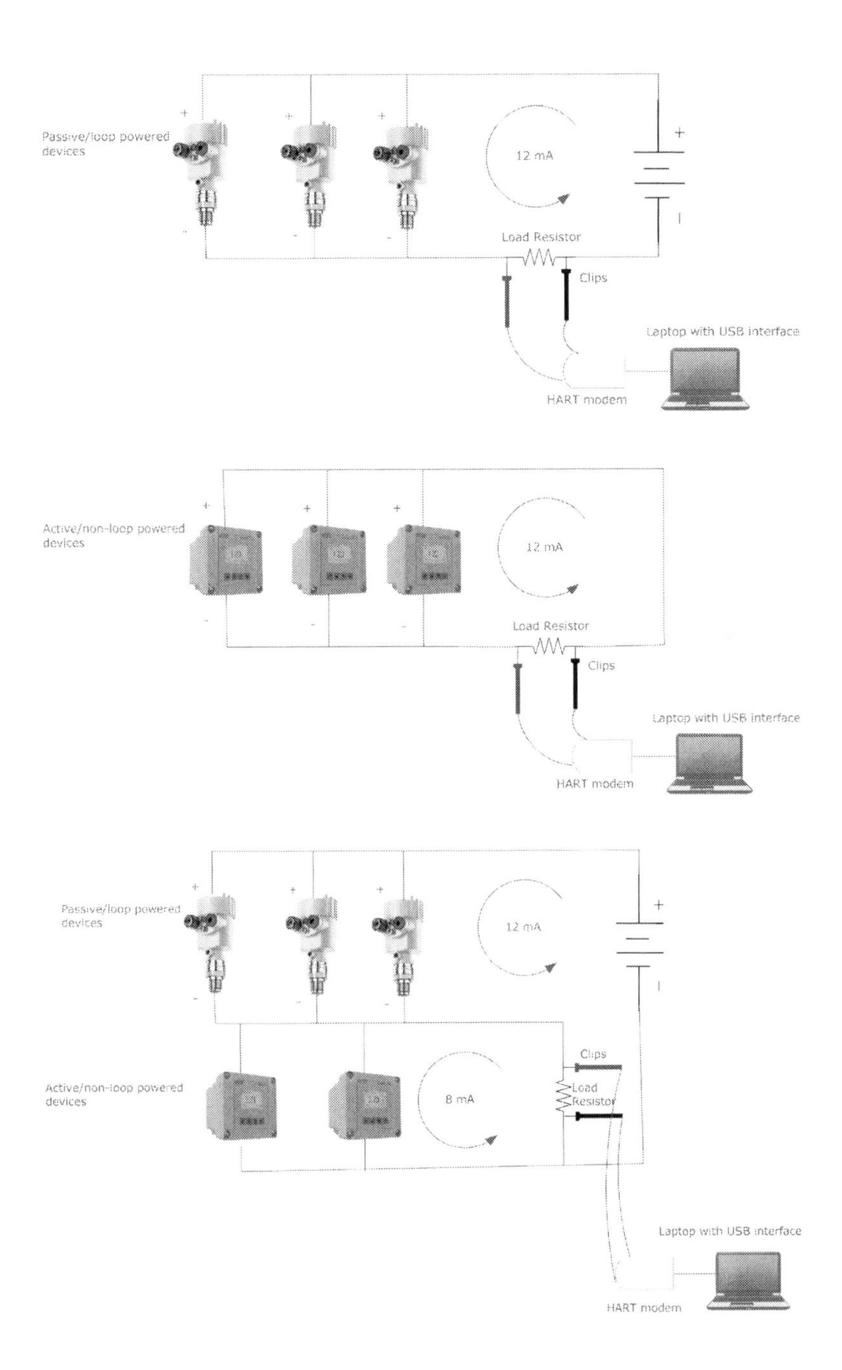
Passive/loop powered devices
12 mA
Load Resistor
Clips
Laptop with USB interface
HART modem
Active/non-loop powered devices
12 mA
Load Resistor
Clips
Laptop with USB interface
HART modem
Passive/loop powered devices
12 mA
Active/non-loop powered devices
8 mA
Clips
Load Resistor
Laptop with USB interface
HART modem

As you can see, calculating the loop resistance and the RC value of the network is going to be different depending on the mix of instruments.

Sizing the HART (or load) resistor is critical. The larger the value, the more noise immunity your network will have. However, you must:

- Ensure that the RC is not too large, or you will not have HART communications.
- Verify that the voltage drop over the HART resistor is not too large during startup. If the voltage drop is too large, then the instruments will not start up.

It may also be difficult to find all the values to do these calculations. In this case, we recommend setting up a sample network initially to measure these values.

HART multidrop... avoid if possible!

I have designed PROFIBUS PA networks and HART multidrop. Give me PROFIBUSPA any day over HART multidrop.

HART multidrop design is not fun, and neither is the wiring.

I know you can save a lot of money and it makes a lot of sense economically if you are using WirelessHART, but it is definitely not enjoyable.

My basic rule of thumb is to keep the distances and number of devices very small. Wire up a test network and see how it works. If it works fine, then you should be good to go. If you can avoid it, then, by all means, do so.

TROUBLESHOOTING

In the many HART sites that I have been involved with, it is quite often the case that problems lie either in something simple or in the network's design (or, more often than not, the lack of design).

Most people who work with HART a lot will have a small list of tricks to try when they cannot establish communications to an instrument. Here is mine:

1. Verify that you have power at the instrument. Simple wiring issues like having the plus and minus reversed are common enough. Sometimes I have seen cases where the instrument powers up but then shuts itself off once the reading starts working.

   This one is caused by too much resistance in the line causing too large a power drop at higher mA, which causes the voltage at the instrument to fall below the minimum value for that instrument. In this case, you can either increase the voltage or find a method for lowering the resistance.

2. Try a different resistance for your HART resistor—either increase the value or lower it and see if it then works.

3. Check to see where the wire is routed. If it is running near a power cable or variable frequency drive, then relocate it.

For point-to-point HART, these three checks work most of the time.

There are also two tools that you can use to help you. First off is an oscilloscope. Looking at the waveforms will let you know if there is other electrical noise at the same frequency as the HART signal that is killing the signal. This is the best tool but is expensive and can be hard to carry around.

Alternatively, you can use a HART speaker to listen for the HART signal. This troubleshooting aid is not available on the market but can be easily made. See Appendix A for details. The HART signal will sound like a chirp. If you don't hear a chirp, then there is no signal. When there is no signal, then troubleshooting becomes an exercise in reviewing installation methods and network design.

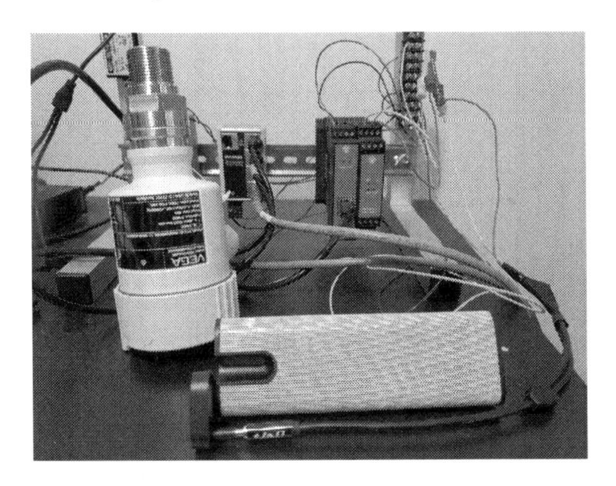

If you hear a chirp, but it does not sound 'normal,' or you hear a lot of static over the chirp, then you have a noise situation. In this case, troubleshooting again becomes an exercise in reviewing installation methods and network design. The HART speaker is low cost and easy to carry around.

CHAPTER 5: HART-IP

*"If you want to make God laugh,
tell him about your plans."*
--Woody Allen

HART over copper and HART-IP both have one implementation issue in common: people don't plan.

Mr. Allen may have been correct about how we humans plan too much and that our plans are not realistic, but with networks, this quote would be better reworded as "If you want to make God laugh, tell him about your *lack* of planning!"

This chapter talks all about HART over Ethernet, HART-IP. As with any network, planning is key.

HART-IP

HART-IP is used as an intermediary between a HART enabled I/O device and the control system and configuration software.

HART-IP encapsulates a HART message into an IP message which can be transported either via TCP or UDP from the server (HART enabled I/O device) and a HART client (control system or configuration software).

For understanding communications systems, there is a model called the OSI 7 layer model that is very useful. The basic idea of the model is to divide the tasks required to communicate from point A to point B into seven different layers. Each layer adds to communication packet and provides a particular function.

For understanding Ethernet-based communications system, this model is very useful. Ethernet is the lower layer and half of layer 1 plus half of layer 2. There are many different protocols that sit on top of Ethernet physical layer. These protocols have come to be know as Ethernet family of protocols.

Some of these protocols do the same thing. For example Transmission Control Protocol (TCP) and User Datagram Protocol (UDP) are both at the Transport level and do the same basic thing but in a different way. Both provide a connection.

TCP provides a secure connection where the data is guaranteed to get to its location. If it does not get there, then retries are done and finally if it still does not get there, then an error is generated. While UDP does not have this guarantee or retries. It provides what is referred to as a connectionless connection.

Using the OSI 7 layer model as a reference, the packet buildup for HART-IP would look like the following:

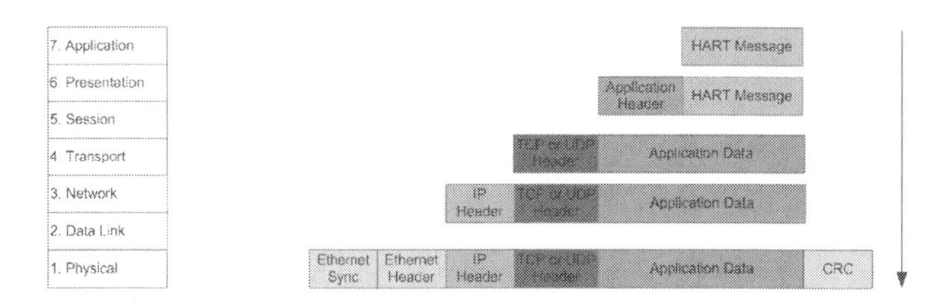

By encapsulating most of the HART message, it makes it easy for the various configuration softwares packages to drill right down to an instrument over an Ethernet network.

The HART-IP to HART gateways also tend to support one of the industrial Ethernet protocols such as Modbus TCP, Ethernet/IP or PROFINET. With any of these protocols, the control system can receive cyclic updates on both the process variable and various status bytes.

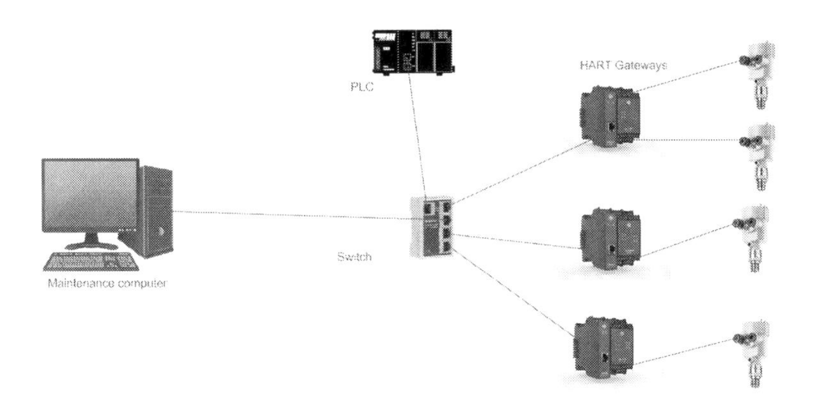

In the network example pictured here, the gateways are by Thorsis Technologies, which supports Modbus TCP or PROFINET. The PLC would use either Modbus TCP or PROFINET to pull the process variables and status information from the gateways.

The gateway then uses HART to pull this information from the instrument. The maintenance computer then uses HART over IP to configure and troubleshoot the field devices through the gateways.

SWITCHES

A switch is a device that takes the message coming in one port and 'routes' it to the port that has the device that the message is for.

Choosing the correct switch for your application is very important. HART over IP has no special requirements for a switch. Therefore, whatever switch you choose for the industrial protocol you are using will work fine with HART over IP.

There are two basic types of switches: a simple, lower-cost unmanaged switch or a managed switch, which will have more configuration and network diagnostic information.

The debate between a managed and unmanaged switch is a difficult one. The cost difference is significant; however, so is the functionality.

Both will work, but later on, when you are trying to maintain a large network, your maintenance staff will wish that you had purchased managed switches.

ROUTERS

A router is used to join one network to another and can also be used to isolate traffic from one area to another. These are very useful devices in your overall network design. However, routers do pose a problem to some industrial protocols such as PROFINET, which cannot pass messages through routers.

HART-IP, on the other hand, is easily able to go through a router. This is because HART-IP uses IP network layer, which contains the routing information. In HART-IP's case, all you have to do is open port 5094 in the router. A port can be thought of as a type of mailbox. Routers have many of the 'mailboxes.' Most of them will be blocked by default, but there will always be a setting to 'open' a particular port. Different protocols will require different ports to be opened. For HART-IP the port is 5094.

By being router-friendly, the maintenance computer can be located anywhere and does not have to be located near the PLC or devices. This opens the door to remote connections where your maintenance person could access the devices from offsite and troubleshoot problems remotely.

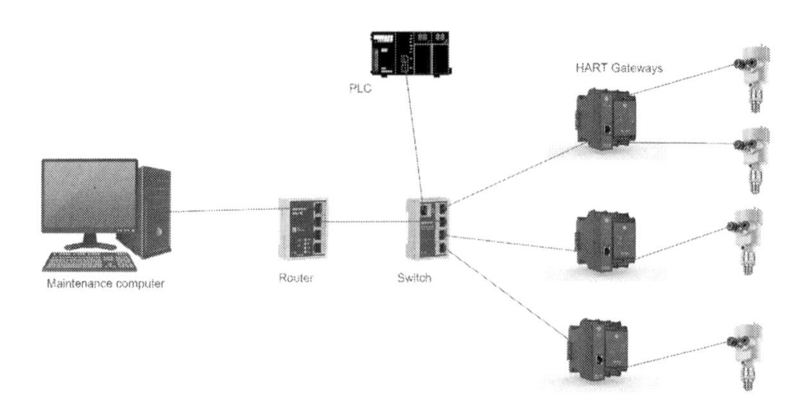

In this topology diagram, the maintenance computer is located away from the automation network on another network, and the HART-IP messages are routed through the router. The router and switch shown here is made by a company called Helmholz from Germany. The gateways are made by Thorsis Technologies from Germany.

NETWORK LOAD AND MONITORING

One interesting fact about industrial Ethernet is that when designing a small system, it is very hard to have it not work properly. However, as the system gets larger, proper network design becomes very important.

Two elements that are sometimes forgotten are the flow of data and the network load. Each device on your network will produce data onto the network, taking up bandwidth. Network load is a measure of this traffic and is often stated as a percent.

If the network load on all lines is under 20%, then there are no issues. If it is between 20% and 50%, then you'd want to watch it and maybe redesign your network to lower this value down to below 20%. If it is over 50% in any location on the network, then your network is in trouble and you need to do something about it.

For PROFINET networks, there are programs and formula provided by PROFIBUS PROFINET International that will calculate this value. However, these calculations do not take into account the non-PROFINET traffic. Calculating these numbers can be difficult.

One way around this is to put on a network monitor that will pull this information from the managed switches and display it. These monitors are the eyes and ears of your network.

One of the best on the market is a program called Osiris which runs on either Atlas or Mercury.

Atlas is for permanent monitoring, and Mercury for temporary monitoring.

Both are made by PROCENTEC, a leader in network diagnostics. Osiris will let you know who is on your network, network-load, lost packets, etc.

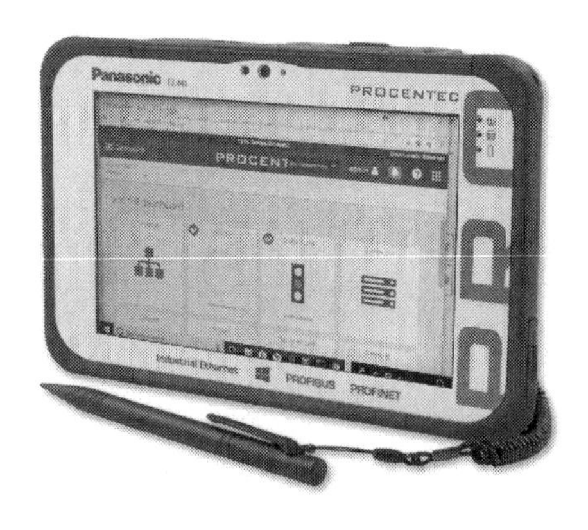

PHYSICAL LAYER

HART IP runs on standard Industrial Ethernet equipment. There are three popular physical layers for Industrial Ethernet:

1. Copper
2. Fiber Optics
3. Wireless

Copper is the most popular and typically runs at 100 Meg bits per second. The wires are a shielded Cat-5e cable that is grounded through the connector at both ends of the run. The wire is simple point to point, where the device is wired into a switch, and then the switch is connected to other switches and the automation system: controller, HMI, configuration software. The runs have to be no more than 100 metres long.

Both fiber optics and wireless have been standardized for Industrial Ethernet. The distances for these physical layers vary according to the fiber used (in the case of fiber optics) and the site layout (in the case of wireless).

Industrial Ethernet is not office Ethernet

I cannot tell you how many times I have gone into an industrial site and seen office-grade Ethernet cables plugged into a control cabinet.

Office-grade Ethernet cables have some noise immunity. However, industrial-grade Ethernet cables have significantly more noise immunity. The industrial-grade cables are shielded and grounded at both ends.

What this means is that if you use office-grade cables in your application, there will be bit errors and, therefore, lost packets. If you are not monitoring your network, you will not notice this, but it will still be affecting your network and over time may even cause dropouts and network outages.

Installing industrial-grade cable into industrial grade switches and following all of the recommended wiring guidelines makes sense and will prevent problems in the future.

DESIGN

A HART-IP network is going to be designed the same as any Industrial Ethernet network: a network of point-to-point connections between devices and switches. The copper cables are limited to a maximum of 100 metres.

The other key element that has already been discussed is making sure that your netload does not get too high.

UPDATE RATE

Ethernet can run at different speeds. Most industrial applications run at 100 Mbps. If HART-IP is running on TCP, then the processing of the TCP part of the message will slow down the actual throughput of the message. However, we are still talking 100 ms update rates.

Since HART-IP will typically be connect to an FSK HART device running at 1200 bps, the HART-IP part of your update rate can be ignored, giving you an overall update rate of around 500 ms.

TROUBLESHOOTING

When troubleshooting WirelessHART, the two most common issues are; I cannot see the gateway and I cannot see the instrument from my control system.

If the control system cannot see the gateway, then you need to investigate the Ethernet network connection between the control system and the gateway. First checking to see that the gateway is powered up. This last question seems so simple, but you would be surprised how many times I have solved 'communication problems' by simply turning on the power.

To check the link from the controller to the gateway, you need to use a computer that can talk over Ethernet to the controller. Then use a very useful command called 'PING.'

From the Windows start menu, run the program CMD which opens a DOS window. From that window, you can type 'PING' followed by a space and then the IP address of the gateway.

This is key for first-level troubleshooting. The command hopefully will return the time it took to get a return message back from that IP address. If you cannot get a ping back, then you have network issues of some sort. For example, you may be on the wrong subnet or router.

To troubleshoot the link between the gateway and the instrument takes us back to troubleshooting standard HART that we covered at the end of chapter 4. The gateway may help in this regard by providing information on what is going on, but this will depend on the gateway.

CHAPTER 6: WIRELESSHART

"Mystery creates wonder and wonder is the basis of man's desire to understand."
--Neil Armstrong

Wireless tends to generate wonder all by itself. It is a mystery, a black art—or at least that is how it has been viewed. In reality, it is pretty easy now and comes down to some key physical elements: line of sight and avoiding obstructions. This chapter will go over this in detail.

None of it is hard, but the devil is in the details. Keep an eye on these details—if you get them right, then WirelessHART is a very easy network to implement.

We will also look under the hood a little bit—just enough to point out that this technology is actually very complex. There are marvels to wonder at, but it is easy to work with!

WIRELESSHART

WirelessHART was added in V7 of the protocol. They wanted the communications to be reliable, secure, and simple, with easy connections of existing HART devices and using all the standard HART services.

WirelessHART was able to meet all of these requirements and more. What is particularly interesting is how the designers were able to hide very complex technology behind a very simple user interface.

WirelessHART uses IEEE 802.15.4 physical layer. It runs at 2.4 GHz and runs on low power radios. The data rates are up to 250 kbit/s. It uses special frequency hopping technology to minimize the impact of noise and other protocols running at 2.4 GHz. To connect devices, it uses time-synchronized mesh technology.

The idea of the mesh is that there are multiple communications paths back to the gateway. The messages are passed from one device to another until they get back to the WirelessHART gateway.

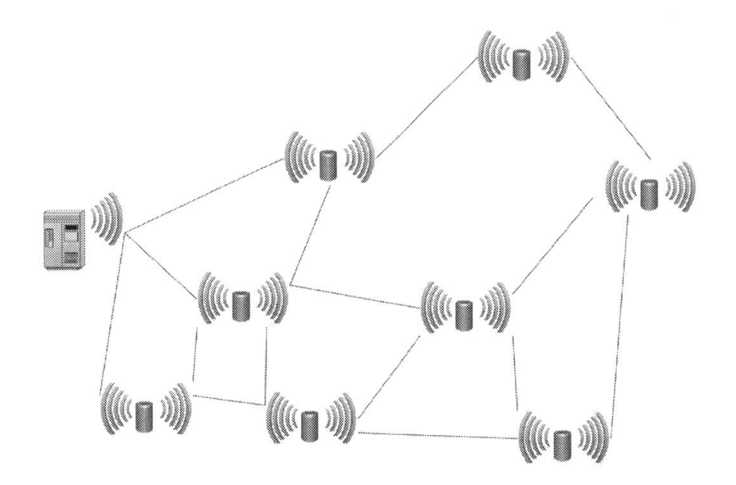

Under the hood, the WirelessHART engine has a lot going on—many algorithms to optimize communications paths and keep everything working well. The end user, however, does not need to know any of the details. They have to lay it out in a manner where every node can see other nodes in the mesh—the more nodes, the better.

DESIGN

The design is fairly simple. The maximum distance between nodes is 250 metres outdoors and 50 metres indoors. However, this distance is going to vary according to how many and types of obstacles and interferences.

The best way to see this is to do a site audit where you take a gateway and some instruments and position them where you want to put instruments and see what the dB readings are. If you are getting good signals, then all is well. Otherwise, look at changing some of the node locations.

Each device should have at least three neighbours that it can see its signal. The device's antenna needs to be mounted more than 0.5 metres from any vertical surface and should be mounted more than 1.5 metres off the ground. Gateways should have at least five neighbours and can support up to 100 devices.

Since obstacles have such a large effect on wireless, the design may need some minor changes once you are in the field doing the installation.

Having some extra repeaters on hand is a good idea. Another best practice is having a good network drawing showing where everything is installed.

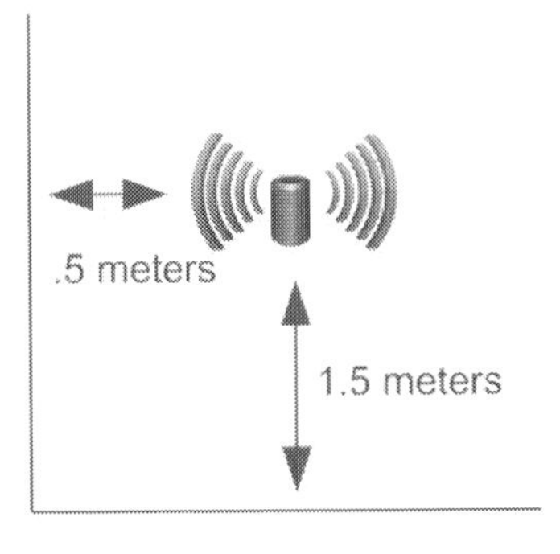

Oh-where-oh-where has my instrument gone!

A network layout is key for troubleshooting any network, but is particular important for WirelessHART. The reason that I say this is because I have been in many plants using wired devices that staff did not know the location of, and we had to find them by tracing wiring.

In the case of WirelessHART there are no wires to trace, so just imagine how much fun it would be to find an instrument whose battery has died! No signal, no wires—good luck!

COMMISSIONING A WIRELESSHART NETWORK

Once you have installed and powered up all the instruments, to have them join the network, you must enter the Network ID and Network Join Key into the instruments. Both of these parameters come from the gateway.

You also have to set the refresh rate—when picking the refresh rate, try to keep it as slow as possible to conserve battery life.

After setting those three parameters in the field device, the next step is to wait for the network to come up. The network coming up might take a while: think minutes and hours, not seconds. The amount of time depends on the different refresh rates and the number of instruments and neighbours. For even a network of 20 to 30 instruments, going for lunch while the network establishes itself is not a bad idea!

The last step is looking in the gateway to verify that all the instruments have joined the network.

TROUBLESHOOTING

For all the complexities hidden in this technology, WirelessHART is remarkably easy to troubleshoot. Most of the issues are installation issues/positioning issues.

Can the instrument get enough signal?Does the instrument have enough neighbours? The diagnostic section of your gateway will show you this information —after that, you need to just work with modifying the positioning of the instruments and the number of repeaters.

Just keep in mind line of sight, obstructions, and distance. Try to have line of sight between neighbouring instruments if you can. Keep obstructions to a minimum, particularly metal ones. The design rules have the maximum recommended distances but really, the shorter, the better.

This holds for both WirelessHART devices and WirelessHART adaptors. WirelessHART adaptors can also have communication issues between the adaptor and the HART slave.

Therefore, for the adaptors, your troubleshooting will be in two parts: first, can I see the adaptor? And then, can the adaptor see the instrument?

CHAPTER 7: WHERE DO I GO FROM HERE?

"Begin at the beginning," the King said gravely, "and go on till you come to the end: then stop."
--Lewis Carroll

We have now come to the end. According to the King, it is time to stop, but I do not think so.

Reading this book is just the beginning. Yes, there is still more to learn. Luckily there are many resources open to you.

People learn by reading, by being taught, and by doing.

READING

Since this was really just an introduction with the focus on working with the protocol, there are many details of the protocol missing from this book.

HART is owned and managed by the FieldComm Group which has many articles on HART on their website at https://fieldcommgroup.org.

There is a detailed book on HART Technology that is an excellent reference book.

The book is called *HART Technology – A Technical Overview* by Romilly Bowden.
ISBN-13: 978-1549862274, ISBN-10: 1549862278.

COURSES

Some HART training is incorporated into many instrumentation courses. However, it is rare to have a course just about HART and how to properly use it.

One such course, "HART, SIMATIC PDM, PACTware Maintenance," is offered by JCOM Automation Inc. in Peterborough, Ontario, Canada. I designed this course and am its main instructor. Learn more about this course at jcomautomation.ca/training.

This course is for those who install and maintain HART installations. The class provides a mixture of theory and hands-on experience to give you the skill set you need, and the ability to put it into practice immediately.

Attendees will learn how to analyze, localize, and solve HART problems. The theory covered includes details about cabling, shielding, and grounding for HART, WirelessHART and HART-IP. The practical hands-on examples show you how to use various tools to find and fix your problems quickly.

DOING

The old saying that experience is the best teacher is very true. For advanced technology like HART, you need some basic information, which this book has provided.

However, the only way to really learn this information is by applying it!

APPENDIX A: HOW TO MAKE A HART SPEAKER

To hear the signal, you will need the following:

- Mini amplified speaker
- 15 K ohm resistor
- 0.1 uF ceramic disc capacitor
- Cable
- Two test hook clips

The wiring diagram* is given below:

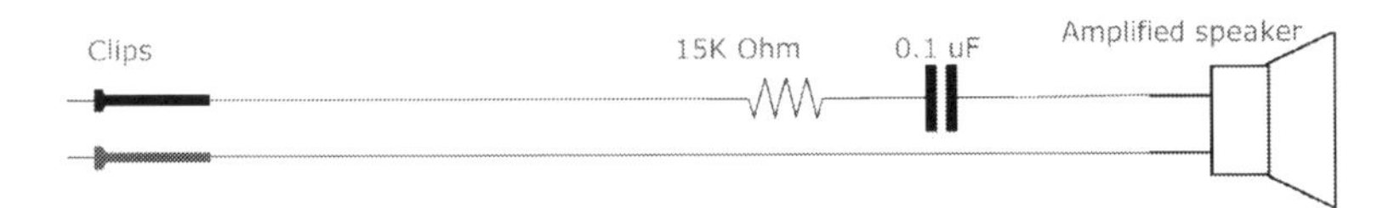

*We provide this drawing as reference and accept no responsibility to its accuracy or any liability to its use.

APPENDIX B: ACRONYMS

CRC: Cyclic Redundancy Check

C8PSK: Coherent 8-way Phase Shift Keying

DAC: Digital to Analog Converter

DD: Device Description

DTM: Device Type Manager

EDD: Electronic Device Description

FDI: Field Device Integration

FDT: Field Device Tool

FSK: Frequency Shift Keying

HART: Highway Addressable Remote Transducer

IP: Internet Protocol

P-P: Peak to Peak

RC: Resistive Capacitance

RMS: Root Mean Square

TCP: Transmission Control Protocol

UDP: User Datagram Protocol

USB: Universal Serial Bus

ABOUT THE AUTHOR

James Edward Powell, P. Eng. is the Principal Engineer and owner of JCOM Automation Inc. in Peterborough, Ontario, Canada.

After escaping Queen's University in Kingston with an Honours BSc. in Mathematics and Engineering – Electrical Controls and Communication option, James left the theoretical world of his degree to the practical world of engineering and making things work.

He found out that troubleshooting was his passion when as a Micro VAX system manager he realized that his most enjoyable days working were the days when something went wrong with the system and he had to spend hours figuring out what was wrong and making it work again.

This love of troubleshooting combined with a great attention for detail lead him into the world of industrial communications. First as an application engineer for GE FANUC, then as an Automation specialist with GE Engineering Services and a product manager with Milltronics/Siemens.

Now, with over 30 years of experience in industrial communications, James has a wealth of knowledge that he enjoys sharing with his students in the many technical training courses that he teaches each year. He loves escaping the office and troubleshooting customers networks and generally solving problems.

James can be contacted at:

JCOM Automation Inc.
1115 Whitefield Dr.
Peterborough, Ontario
Canada, K9J-7P4
jamesp@jcomautomation.ca
jcomautomation.ca
Phone: +1-705-868-8745

OTHER BOOKS BY JAMES POWELL

Catching the Process Fieldbus – An introduction to PROFIBUS and PROFINET by James Powell, P. Eng., and Henry Vandelinde, Ph.D.

In March 2009, James Powell, along with Dr. Henry Vandelinde of Siemens, released an introductory level book on PROFIBUS called Catching the Process Fieldbus.

This book has been translated into German, Spanish and Chinese. In 2015, they released the 2nd edition of this book that expanded it to include PROFINET.

The PDF version of this book is available for free download at jcomautomation.ca/books.

Printed in Poland
by Amazon Fulfillment
Poland Sp. z o.o., Wrocław